TECH MANUAL TO ACCOMPANY

COLLISION REPAIR AND REFINISHING

A FOUNDATION COURSE FOR TECHNICIANS

Alfred M. Thomas

DELMAR
CENGAGE Learning

Australia • Brazil • Japan • Korea • Mexico • Singapore • Spain • United Kingdom • United States

DELMAR
CENGAGE Learning

Tech Manual to Accompany
Collision Repair and Refinishing
A Foundation Course for Technicians
Alfred M. Thomas

Director of Learning Solutions:
Sandy Clark

Executive Editor:
David Boelio

Senior Product Manager:
Matthew Thouin

Editorial Assistant:
Jillian Borden

Vice President, Career and
Professional Marketing: Jennifer McAvey

Executive Marketing Manager:
Deborah S. Yarnell

Marketing Manager:
Jimmy Stephens

Marketing Coordinator:
Mark Pierro

Production Director:
Wendy Troeger

Production Manager:
Mark Bernard

Content Project Manager:
Cheri Plasse

Art Director:
Benj Gleeksman

For product information and technology assistance, contact us at
Cengage Learning Customer & Sales Support, 1-800-354-9706
For permission to use material from this text or product,
submit all requests online at **cengage.com/permissions**
Further permissions questions can be emailed to
permissionrequest@cengage.com

Library of Congress Control Number: 2007941007

ISBN-13: 978-1418001025

ISBN-10: 1418001023

Delmar
5 Maxwell Drive-2919
Clifton Park, NY 12065
USA

Cengage Learning is a leading provider of customized learning solutions with office locations around the globe, including Singapore, the United Kingdom, Australia, Mexico, Brazil, and Japan. Locate your local office at:
international.cengage.com/region

Cengage Learning products are represented in Canada by Nelson Education, Ltd.

For your lifelong learning solutions, visit **delmar.cengage.com**

Visit our corporate website at **www.cengage.com**

Notice to the Reader
Publisher does not warrant or guarantee any of the products described herein or perform any independent analysis in connection with any of the product information contained herein. Publisher does not assume, and expressly disclaims, any obligation to obtain and include information other than that provided to it by the manufacturer. The reader is expressly warned to consider and adopt all safety precautions that might be indicated by the activities described herein and to avoid all potential hazards. By following the instructions contained herein, the reader willingly assumes all risks in connection with such instructions. The publisher makes no representations or warranties of any kind, including but not limited to, the warranties of fitness for particular purpose or merchantability, nor are any such representations implied with respect to the material set forth herein, and the publisher takes no responsibility with respect to such material. The publisher shall not be liable for any special, consequential, or exemplary damages resulting, in whole or part, from the readers' use of, or reliance upon, this material.

Printed in the United States of America
1 2 3 4 5 XX 11 10 09

CONTENTS

IV   Contents

PREFACE

The *Tech Manual to accompany Collision Repair & Refinishing* is designed to work hand-in-hand with the textbook to offer additional opportunities for review and application of the material covered in the book's respective chapters. The *Tech Manual* includes the following chapter components:

Work Assignments: These are theory-based assignments that should be completed following a thorough study of the chapter content.

Work Orders: The Work Orders are procedure-based activities designed to direct live tasks in the shop in conjunction with demonstration and guidance from an instructor.

Review Questions: The questions at the end of each *Tech Manual* chapter will help reinforce the above activities. The review questions section could be assigned to students as a quiz, in-class assignment, or for homework.

Chapter 1

Introduction to Collision Repair

■ WORK ASSIGNMENT 1-1

COLLISION REPAIR OPERATION

Name _____   Date _____

Class _____   Instructor _____   Grade _____

1. After reading the assignment, in the space provided below, list the personal and environmental safety equipment and precautions needed for this assignment.

   _____

   _____

   _____

2. What purpose does an estimate serve? Record your findings:

   _____

   _____

   _____

3. How do an estimate and a work order differ? Record your findings:

   _____

   _____

   _____

4. What part does an insurance company play in the repair process? Record your findings:

   _____

   _____

   _____

5. What is meant by a DRP? Record your findings:

   _____

   _____

   _____

6. What is a supplemental estimate? Record your findings:

   _____

   _____

   _____

7. What is a damage report? Record your findings:

   _____

   _____

   _____

8. A technician who works in the "metal works" part of the shop does what type of work? Record your findings:

_____

_____

_____

9. A technician who works in the "structural repair" part of the shop does what type of work? Record your findings:

_____

_____

_____

10. A technician who works in the "mechanical/electrical" part of the shop does what type of work? Record your findings:

_____

_____

_____

11. A technician who works in the "refinish" part of the shop does what type of work? Record your findings:

_____

_____

_____

12. Why is vehicle information such as a VIN and a service part ID label important to the repair process? Record your findings:

_____

_____

_____

INSTRUCTOR COMMENTS:

_____

_____

_____

_____

_____

_____

CAREER OPPORTUNITIES IN COLLISION REPAIR

Name _____ Date _____

Class _____ Instructor _____ Grade _____

1. After reading the assignment, in the space provided below, list the personal and environmental safety equipment and precautions needed for this assignment.

   _____

   _____

   _____

2. Describe your feelings regarding the opportunities in the collision repair industry. Record your findings:

   _____

   _____

3. Describe what a metal repair technician does. Record your findings:

   _____

   _____

   _____

4. Describe what a structural repair technician does. Record your findings:

   _____

   _____

5. Describe what a refinish technician does. Record your findings:

   _____

   _____

6. Describe what a detailing technician does. Record your findings:

   _____

   _____

7. Describe what a helper/apprentice does. Record your findings:

   _____

   _____

8. Describe what a collision estimator does. Record your findings:

   _____

   _____

   _____

9. Describe what a production manager does. Record your findings:

   _____

   _____

   _____

10. Describe what a parts manager does. Record your findings:

_____

_____

_____

11. List other related jobs that the collision repair industry may have. Record your findings:

_____

_____

_____

12. In 10 years, which job would you like to be doing in the collision repair industry? Record your findings:

_____

_____

_____

INSTRUCTOR COMMENTS:

_____

_____

_____

_____

_____

_____

WHAT IS COLLISION REPAIR?

Name _____   Date _____

Class _____   Instructor _____   Grade _____

OBJECTIVES

- Identify the various types of collision repair centers, including:
  - Independent
  - Dealership
  - Franchise
- Recognize the basics of collision repair operations, including:
  - Estimating collision damages
  - Metal work
  - Structural repairs
  - Mechanical/Electrical
  - Refinishing
  - Vehicular identification
  - Service information retrieval
  - Post-paint operations
- Distinguish between collision repair career opportunities, including:
  - Collision repair technician
  - Collision repair shop owner
  - Damage estimator
  - Production manager
  - Parts manager
  - Office manager
  - Receptionist
  - Related collision industry careers
  - Insurance adjuster or appraiser

NATEF TASK CORRELATION

The written and hands-on activities in this chapter satisfy the NATEF High Priority-Individual and High Priority-Group requirements. Though there are no NATEF requirements for this section, it is felt that a working knowledge of the industry is necessary.

Tools and Equipment needed (NATEF tool list)

- Pen and pencil
- Safety glasses
- Gloves

Instructions

The activities for this section are intended to acquaint you with the shop and its equipment. Your instructor will train each of you on the safe use of the equipment that you will identify today. Safety is a primary concern and you must follow all instructions, both written and demonstrated, and should not operate equipment without the expressed direction of your instructor.

PROCEDURE

In the lab, after completing the work assignment for that section, follow the work order and record your findings for your instructor to review. If the section calls for "instructor approval," you should not proceed without this approval.

1. After reading the work order, gather the safety gear needed to complete the task. In the space provided below, list the personal and environmental safety equipment and precautions needed for this assignment. Have the instructor check and approve your plan before proceeding.

_____

_____

_____

INSTRUCTOR'S APPROVAL _____

2. With the list provided by your instructor, find the location of the safety equipment on the list and document their location.

Fire extinguishers:

_____

_____

_____

Eyewash stations:

_____

_____

_____

First-aid kit:

_____

_____

_____

Exits:

_____

_____

_____

Fire alarm:

_____

_____

_____

Fire blankets:

_____

_____

_____

Storm shelter:

_____

_____

_____

Electrical shutoff switch:

_____

_____

_____

Phone:

_____

_____

_____

Light switch:

_____

_____

_____

INSTRUCTOR COMMENTS:

_____

_____

_____

_____

_____

_____

COLLISION REPAIR OPERATION

Name _____   Date _____

Class _____   Instructor _____   Grade _____

OBJECTIVES

- Identify the various types of collision repair centers, including:
  - Independent
  - Dealership
  - Franchise
- Recognize the basics of collision repair operations, including:
  - Estimating collision damages
  - Metal work
  - Structural repairs
  - Mechanical/Electrical
  - Refinishing
  - Vehicular identification
  - Service information retrieval
  - Post-paint operations
- Distinguish between collision repair career opportunities, including:
  - Collision repair technician
  - Collision repair shop owner
  - Damage estimator
  - Production manager
  - Parts manager
  - Office manager
  - Receptionist
  - Related collision industry careers
  - Insurance adjuster or appraiser

NATEF TASK CORRELATION

The written and hands-on activities in this chapter satisfy the NATEF High Priority-Individual and High Priority-Group requirements. Though there are no NATEF requirements for this section, it is felt that a working knowledge of the industry is necessary.

Tools and Equipment needed (NATEF tool list)
- Pen and pencil
- Safety glasses
- Gloves

Instructions

The activities for this section are intended to acquaint you with the shop and its equipment. Your instructor will train each of you on the safe use of the equipment that you will identify today. Safety is a primary concern and you must follow all instructions, both written and demonstrated, and should not operate equipment without the expressed direction of your instructor.

PROCEDURE

In the lab, after completing the work assignment for that section, follow the work order and record your findings for your instructor to review. If the section calls for "instructor approval," you should not proceed without this approval.

1. After reading the work order, gather the safety gear needed to complete the task. In the space provided below, list the personal and environmental safety equipment and precautions needed for this assignment. Have the instructor check and approve your plan before proceeding.

    _____
    _____
    _____
    _____

INSTRUCTOR'S APPROVAL _____

2. In the shop locate the estimating area and document its location. Notes:

    _____
    _____
    _____

3. In the shop locate the metal work area and document its location. Notes:

    _____
    _____
    _____

4. In the shop locate the structural repair area and document its location. Notes:

    _____
    _____
    _____

5. In the shop locate the mechanical/electrical area and document its location. Notes:

    _____
    _____
    _____

6. In the shop locate the refinish area and document its location. Notes:

    _____
    _____
    _____

7. In the shop locate the tool storage area and document its location. Notes:

    _____
    _____
    _____

8. In the shop locate the supply area and document its location. Notes:

    _____
    _____
    _____

9. On a vehicle locate its VIN and record it. Notes:

    _____
    _____
    _____

10. On a vehicle find the manufacturer's paint code and record it. Notes:

_____

_____

_____

INSTRUCTOR COMMENTS:

_____

_____

_____

_____

_____

_____

CAREER OPPORTUNITIES IN COLLISION REPAIR

Name _____   Date _____

Class _____   Instructor _____   Grade _____

OBJECTIVES

- Identify the various types of collision repair centers, including:
  - Independent
  - Dealership
  - Franchise
- Recognize the basics of collision repair operations, including:
  - Estimating collision damages
  - Metal work
  - Structural repairs
  - Mechanical/Electrical
  - Refinishing
  - Vehicular identification
  - Service information retrieval
  - Post-paint operations
- Distinguish between collision repair career opportunities, including:
  - Collision repair technician
  - Collision repair shop owner
  - Damage estimator
  - Production manager
  - Parts manager
  - Office manager
  - Receptionist
  - Related collision industry careers
  - Insurance adjuster or appraiser

NATEF TASK CORRELATION

The written and hands-on activities in this chapter satisfy the NATEF High Priority-Individual and High Priority-Group requirements. Though there are no NATEF requirements for this section, it is felt that a working knowledge of the industry is necessary.

Tools and Equipment needed (NATEF tool list)

- Pen and pencil
- Safety glasses
- Gloves

Instructions

The activities for this section are intended to acquaint you with the shop and its equipment. Your instructor will train each of you on the safe use of the equipment that you will identify today. Safety is a primary concern and you must follow all instructions, both written and demonstrated, and should not operate equipment without the expressed direction of your instructor.

PROCEDURE

In the lab, after completing the work assignment for that section, follow the work order and record your findings for your instructor to review. If the section calls for "instructor approval," you should not proceed without this approval.

1. After reading the work order, gather the safety gear needed to complete the task. In the space provided below, list the personal and environmental safety equipment and precautions needed for this assignment. Have the instructor check and approve your plan before proceeding.

_____

_____

_____

INSTRUCTOR'S APPROVAL _____

2. Locate the detailing area and document its location. Notes:

_____

_____

3. Locate where the detailing supplies are kept and document their locations. Notes:

_____

_____

4. Locate where the welding equipment is stored and document its location. Notes:

_____

_____

5. Locate where the refinish supplies are stored and document their locations. Notes:

_____

_____

6. Locate where the estimates and work orders are stored and document their locations. Notes:

_____

_____

7. Locate where the cleanup equipment is stored and document its location. Notes:

_____

_____

8. Locate where the trash containers are and where they are to be emptied and document their locations. Notes:

_____

_____

9. Locate the restroom and document its location. Notes:

_____

_____

_____

INSTRUCTOR COMMENTS:

_____
_____
_____
_____
_____
_____

Name _____   Date _____

Class _____   Instructor _____   Grade _____

1. All of the following are operations associated with collision repair EXCEPT:
   A. creating a damage report
   B. vacuuming the interior carpets
   C. manufacturing a new fender
   D. retrieving and clearing DTCs

2. *Technician A* says that a collision repair technician must be concerned with durability and performance. *Technician B* says that a collision repair technician is only concerned with structural integrity. Who is correct?
   A. Technician A only
   B. Technician B only
   C. Both Technicians A and B
   D. Neither Technician A nor B

3. In addition to restoring a collision-damaged vehicle to preloss condition, a collision repair shop seeks to restore:
   A. the owner's deductible
   B. the owner's confidence in the vehicle
   C. the insurance company's subrogation payments
   D. the shop's reputation in the industry

4. **TRUE** or **FALSE**: A collision repair center can be dealership owned and operated.

5. *Technician A* says that a direct repair program rarely, if ever, involves an insurance company. *Technician B* says that a combo repairer may work on a collision-damaged vehicle from start to finish. Who is correct?
   A. Technician A only
   B. Technician B only
   C. Both Technicians A and B
   D. Neither Technician A nor B

6. The collision repair process begins with:
   A. a work input order
   B. a work order
   C. a detailing process
   D. a damage analysis

7. All of the following are parts of the refinishing process EXCEPT:
   A. bleeding
   B. sealing
   C. sanding
   D. corrosion protection

8. *Technician A* says that a road test is rarely necessary on collision-repaired vehicles. *Technician B* says that most shops do their own wheel alignments. Who is correct?
   A. Technician A only
   B. Technician B only
   C. Both Technicians A and B
   D. Neither Technician A nor B

9. *Technician A* says that a crash book usually contains everything needed to create an estimate of damages on a vehicle. *Technician B* says digital imaging can be utilized by collision repairers and insurers to correspond with each other. Who is correct?
   A. Technician A only
   B. Technician B only
   C. Both Technicians A and B
   D. Neither Technician A nor B

10. Which of the following is a repair method that a typical metal technician would utilize?
    A. shrink tubing
    B. hammer-on, hammer-off
    C. sound deadening
    D. hammer-up, hammer-down

11. Which of the following is NOT a vehicular frame style?
    A. unibody
    B. ladder
    C. inner body
    D. space frame

12. A late-model four-door GM vehicle is brought into a collision repair center. It is found to have a closed gap between the left fender and front door, a misaligned hood, a cracked windshield with no evidence of anything having struck it, and a buckle in the roof. The vehicle most probably has:
    A. structural misalignment
    B. mechanical damage
    C. non-OEM parts
    D. a faulty collision avoidance system

13. *Technician A* says that direct damages are the result of energy having traveled through the vehicle's structure. *Technician B* says that indirect damages are found in the area of the direct impact to the vehicle. Who is correct?
    A. Technician A only
    B. Technician B only
    C. Both Technicians A and B
    D. Neither Technician A nor B

14 . *Technician A* says that he is a metal repair technician who utilizes hammers, spoons, picks, and prybars. *Technician B* says that she is a refinish technician and has painted aluminum, high-strength steel, and plastic parts on vehicles. Who is correct?
    A. Technician A only
    B. Technician B only
    C. Both Technicians A and B
    D. Neither Technician A nor B

15. **TRUE** or **FALSE**: The VIN plate, located on the top of the dash pad, can give a refinish technician the paint code and color.

Chapter 2

Collision Repair Safety

■ WORK ASSIGNMENT 2-1

REGULATIONS, REGULATORS, AND MSDS

Name _____   Date _____

Class _____   Instructor _____   Grade _____

NATEF TASK IV A. 1-6

1. After reading the assignment, in the space provided below, list the personal and environmental safety equipment and precautions needed for this assignment.

   _____

   _____

   _____

2. How does a positive safety attitude benefit workers and the shop? Record your findings:

   _____

   _____

   _____

3. What roles do the three governing bodies control?
   OSHA:

   _____

   _____

   _____

   EPA:

   _____

   _____

   _____

   DOT:

   _____

   _____

   _____

4. List and explain the four characteristics that would make a substance hazardous. Record your findings:

   _____

   _____

   _____

5. Explain what each of the following anachronisms are and what they represent?

CAAA:

_____

_____

_____

RCRA:

_____

_____

_____

PPE:

_____

_____

_____

MSDS:

_____

_____

_____

INSTRUCTOR COMMENTS:

_____

_____

_____

_____

_____

_____

HAZARDOUS MATERIALS, HAZARDOUS WASTE, AND PERSONAL PROTECTIVE EQUIPMENT (PPE)

Name _____   Date _____

Class _____   Instructor _____   Grade _____

NATEF TASK IV A. 1-6

1. After reading the assignment, in the space provided below, list the personal and environmental safety equipment and precautions needed for this assignment.

   _____

   _____

   _____

2. Though most MSDSs have from 12 to 16 sections, only 8 are required. List the required eight and give a short explanation of each.

   Section 1:

   _____

   _____

   Section 2:

   _____

   _____

   Section 3:

   _____

   _____

   Section 4:

   _____

   _____

   Section 5:

   _____

   _____

   Section 6:

   _____

   _____

   Section 7:

   _____

   _____

Section 8:

_____
_____
_____

3. Describe the difference between hazardous material and hazardous waste. Record your findings:

_____
_____
_____

4. What relation does Section 8 of an MSDS have with a PPE? Record your findings:

_____
_____
_____

INSTRUCTOR COMMENTS:

_____
_____
_____
_____
_____
_____
_____

STORAGE, SPILLS, FIRE, AND WORKING SAFELY

Name _____   Date _____

Class _____   Instructor _____   Grade _____

NATEF TASK IV A. 1-6

1. After reading the assignment, in the space provided below, list the personal and environmental safety equipment and precautions needed for this assignment.

   _____

   _____

   _____

2. Describe what an NFPA label is and how it is related to PPE? Record your findings:

   _____

   _____

   _____

3. List what you believe to be are the top five personal safety considerations. Record your findings:

   _____

   _____

   _____

4. How could hazardous chemicals enter a worker's body? Record your findings:

   _____

   _____

   _____

5. Which is the most common hazardous chemical? Why? Record your findings:

   _____

   _____

   _____

6. Explain acute exposure. Record your findings:

   _____

   _____

   _____

7. Explain chronic exposure. Record your findings:

   _____

   _____

   _____

8. Explain the difference between air purifying respirators and supply air respirators. Record your findings:

   _____

   _____

   _____

9. What are a positive and a negative seal test? Record your findings:

   _____

   _____

   _____

10. Why are gloves a critical part of a technician's PPE? Record your findings:

_____

_____

_____

11. What steps should be taken if a chemical spill occurs? Record your findings:

_____

_____

_____

12. What is lockout/tagout? Record your findings:

_____

_____

_____

INSTRUCTOR COMMENTS:

_____

_____

_____

_____

_____

_____

REGULATIONS, REGULATORS, AND MSDS

Name _____   Date _____

Class _____   Instructor _____   Grade _____

OBJECTIVES

- Understand hazard communication and employee right-to-know regulations.
- Understand personal safety concerning proper dress, including proper protective eyewear, clothing, gloves, and respiratory and hearing protection.
- Follow safety practices when using personal hand tools, power tools, and equipment.
- Adopt commonsense safety practices that must be a part of using any force application, lifting, and straightening equipment.
- Know the proper handling and storage of all hazardous materials and waste materials generated and the regulations governing their use in the shop.

NATEF TASK CORRELATION

The written and hands-on activity in this chapter satisfy the NATEF High Priority-Individual and High Priority-Group requirements for Section IV: Paint and Refinishing, Subsection A. 1-6.

This section not only deals with paint hazards, it also outlines the activities that a technician should observe throughout the shop!

Tools and equipment needed (NATEF tool list)

- Pen and paper
- Respirator
- Paint suit
- MSDS
- Safety glasses
- Gloves
- Ear protection
- Technical data sheets for paint products

Instructions

The activities for this section are intended to acquaint you with shop safety and its equipment. Your instructor will train each of you on the safe use of the equipment and the personal protective equipment (PPE) that will be needed. Safety is a primary concern and you must follow all instructions, both written (MSDS) and demonstrated, and should not operate equipment without the expressed direction of your instructor.

PROCEDURE

In the lab, after completing the work assignment for that section, follow the work order and record your findings for your instructor to review. If the section calls for "instructor approval" you should not proceed without this approval.

1. After reading the work order, gather the safety gear needed to complete the task. In the space provided below, list the personal and environmental safety equipment and precautions needed for this assignment. Have the instructor check and approve your plan before proceeding.

_____

_____

_____

_____

INSTRUCTOR'S APPROVAL _____

2. Pick five products commonly used by a collision repair technician and locate the correct MSDS. Determine the hazards that each product presents and the PPE required for their use. Record your findings:

_____

_____

_____

3. Using the materials commonly used in the shop area, make a list of at least two products that fall into each of the four hazardous waste categories. Record your findings:

_____

_____

_____

4. Draw a scale model floor plan of the shop area. Identify the location of all the fire extinguishers and identify the class of each extinguisher. Also indicate the location of the emergency fire pulls to sound the alarm in an emergency. Record your findings:

_____

_____

_____

5. Develop a fire escape plan for the shop, including a meeting area outside. Record your findings:

_____

_____

_____

6. Develop a storm safety plan and diagram the route that people should take to get to the protective location for safety. Record your findings:

_____

_____

_____

7. Draw a diagram of the electrical disconnect switches in the shop. Record your findings:

_____

_____

_____

INSTRUCTOR COMMENTS:

_____

_____

_____

_____

_____

_____

HAZARDOUS MATERIALS, HAZARDOUS WASTE, AND PERSONAL PROTECTIVE EQUIPMENT (PPE)

Name _____ Date _____

Class _____ Instructor _____ Grade _____

OBJECTIVES

- Understand hazard communication and employee right-to-know regulations.
- Understand personal safety concerning proper dress, including proper protective eyewear, clothing, gloves, and respiratory and hearing protection.
- Follow safety practices when using personal hand tools, power tools, and equipment.
- Adopt commonsense safety practices that must be a part of using any force application, lifting, and straightening equipment.
- Know the proper handling and storage of all hazardous materials and waste materials generated and the regulations governing their use in the shop.

NATEF TASK CORRELATION

The written and hands-on activities in this chapter satisfy the NATEF High Priority-Individual and High Priority-Group requirements for Section IV: Paint and Refinishing, Subsection A. 1-6.

This section not only deals with paint hazards, it also outlines the activities that a technician should observe throughout the shop!

Tools and equipment needed (NATEF tool list)

- Pen and paper
- Safety glasses
- Respirator
- Gloves
- Paint suit
- Ear protection
- MSDS
- Technical data sheets for paint products

Instructions

The activities for this section are intended to acquaint you with shop safety and its equipment. Your instructor will train each of you on the safe use of the equipment and the personal protective equipment (PPE) that will be needed. Safety is a primary concern and you must follow all instructions, both written (MSDS) and demonstrated, and should not operate equipment without the expressed direction of your instructor.

PROCEDURE

In the lab, after completing the work assignment for that section, follow the work order and record your findings for your instructor to review. If the section calls for "instructor approval," you should not proceed without this approval.

1. After reading the work order, gather the safety gear needed to complete the task. In the space provided below, list the personal and environmental safety equipment and precautions needed for this assignment. Have the instructor check and approve your plan before proceeding.

_____

_____

_____

INSTRUCTOR'S APPROVAL _____

2. Pick five products commonly used by a collision repair technician and locate the correct MSDS. Determine the hazards that each product presents and the PPE required for their use. Record your findings:

_____

_____

_____

3. Demonstrate that your safety glasses meet OSHA standards. Notes:

_____

_____

_____

4. Find an NFPA-labeled product and intermit the label. Notes:

_____

_____

_____

5. Find the MSDS for spraying clear and gather the PPE needed.

_____

_____

_____

6. Locate a hazardous material and explain how it should be stored. What section of the MSDS did you find that information?

_____

_____

_____

7. How should waste coolant be disposed of, and what section of the MSDS did you find the information? Notes:

_____

_____

_____

INSTRUCTOR COMMENTS:

_____

_____

_____

_____

_____

_____

■ WORK ORDER 2-3

STORAGE, SPILLS, FIRE, AND WORKING SAFELY

Name _____   Date _____

Class _____   Instructor _____   Grade _____

OBJECTIVES

- Understand hazard communication and employee right-to-know regulations
- Understand personal safety concerning proper dress, including proper protective eyewear, clothing, gloves, and respiratory and hearing protection.
- Follow safety practices when using personal hand tools, power tools, and equipment.
- Adopt commonsense safety practices that must be a part of using any force application, lifting, and straightening equipment.
- Know the proper handling and storage of all hazardous materials and waste materials generated and the regulations governing their use in the shop.

NATEF TASK CORRELATION

The written and hands-on activities in this chapter satisfy the NATEF High Priority-Individual and High Priority-Group requirements for Section IV: Paint and Refinishing, Subsection A. 1-6.

This section not only deals with paint hazards, it also outlines the activities that a technician should observe throughout the shop!

Tools and equipment needed (NATEF tool list)

- Pen and paper
- Safety glasses
- Respirator
- Gloves

- Paint suit
- Ear protection
- MSDS
- Technical data sheets for paint products

Instructions

The activities for this section are intended to acquaint you with shop safety and its equipment. Your instructor will train each of you on the safe use of the equipment and the personal protective equipment (PPE) that will be needed. Safety is a primary concern and you must follow all instructions, both written (MSDS) and demonstrated, and should not operate equipment without the expressed direction of your instructor.

PROCEDURE

In the lab, after completing the work assignment for that section, follow the work order and record your findings for your instructor to review. If the section calls for "instructor approval," you should not proceed without this approval.

1. After reading the work order, gather the safety gear needed to complete the task. In the space provided below, list the personal and environmental safety equipment and precautions needed for this assignment. Have the instructor check and approve your plan before proceeding.

_____

_____

_____

_____

INSTRUCTOR'S APPROVAL _____

2. Pick five products commonly used by a collision repair technician and locate the correct MSDS. Determine the hazards that each product presents and the PPE required for their use. Record your findings:

_____

_____

_____

3. Locate the spill kit and demonstrate how to contain and clean a 1-gallon reducer spill. Record your findings:

_____

_____

_____

4. There are four different types of gloves available. Make the proper choice for PPE when:
Mixing paint:

_____

_____

_____

Cleaning a spray gun:

_____

_____

_____

Straightening a frame:

_____

_____

_____

Welding:

_____

_____

_____

5. Perform a negative and a positive seal test.

_____

_____

_____

6. Locate and prepare a supply air respirator.

_____

_____

_____

7. A technician brings an open water bottle into the mixing room. How could hazardous chemicals get into that technician's system? Record your findings:

_____

_____

_____

INSTRUCTOR COMMENTS:

_____

_____

_____

_____

_____

Name _____ Date _____

Class _____ Instructor _____ Grade _____

1. *Technician A* says that the third prong on an electrical plug provides extra current during a power surge. *Technician B* says that the third prong is a safety ground. Who is correct?
   A. Technician A only
   B. Technician B only
   C. Both Technicians A and B
   D. Neither Technician A nor B

2. *Technician A* says that solvent-soaked towels should be disposed of in an airtight container. *Technician B* says that floor dry should be disposed of in the same container used for contaminated sweepings. Who is correct?
   A. Technician A only
   B. Technician B only
   C. Both Technicians A and B
   D. Neither Technician A nor B

3. *Technician A* says that some fumes are heavier than air. *Technician B* says that paint storage cabinets should be vented at the top and bottom. Who is correct?
   A. Technician A only
   B. Technician B only
   C. Both Technicians A and B
   D. Neither Technician A nor B

4. *Technician A* says that one must be trained in its use before using an air purifying respirator. *Technician B* says that with proper ventilation, a dust/mist respirator can be used for painting operations. Who is correct?
   A. Technician A only
   B. Technician B only
   C. Both Technicians A and B
   D. Neither Technician A nor B

5. *Technician A* says that the quickest way to distinguish the difference between fire extinguishers is to look at the outlet nozzle. *Technician B* says that a class "ABC" fire extinguisher is filled with carbon dioxide. Who is correct?
   A. Technician A only
   B. Technician B only
   C. Both Technicians A and B
   D. Neither Technician A nor B

6. Solvents are accidentally spilled on the floor and are on fire. *Technician A* says that a Class A fire extinguisher should be used to extinguish the blaze. *Technician B* says that a class ABC extinguisher should be used. Who is correct?
   A. Technician A only
   B. Technician B only
   C. Both Technicians A and B
   D. Neither Technician A nor B

7. *Technician A* says that a secondary container label must be attached to any container that does not have the manufacturer's label on it. *Technician B* says that a secondary container label is not necessary as long as everyone in the shop is aware of the contents in each specified container and it remains in the same location at all times. Who is correct?
   A. Technician A only
   B. Technician B only
   C. Both Technicians A and B
   D. Neither Technician A nor B

8. *Technician A* says that a solvent-filled container brought into the shop area must have a label properly identifying its contents. *Technician B* says that a secondary container label is necessary to mark a container used to store hazardous waste materials. Who is correct?
   A. Technician A only
   B. Technician B only
   C. Both Technicians A and B
   D. Neither Technician A nor B

9. *Technician A* says that OSHA is responsible for protecting the worker in the workplace. *Technician B* says that the EPA's and OSHA's jurisdictions frequently overlap each other. Who is correct?
   - **A.** Technician A only
   - **B.** Technician B only
   - **C.** Both Technicians A and B
   - **D.** Neither Technician A nor B

10. *Technician A* says that before using a chemical with which you are not familiar, you should first read the product usage bulletin. *Technician B* says that you should first consult the product MSDS. Who is correct?
    - **A.** Technician A only
    - **B.** Technician B only
    - **C.** Both Technicians A and B
    - **D.** Neither Technician A nor B

11. *Technician A* says that the CFR is a set of regulations and guidelines used to create a safe work environment. *Technician B* says that the rules and regulations outlined in the CFR come from many of the legislative acts. Who is correct?
    - **A.** Technician A only
    - **B.** Technician B only
    - **C.** Both Technicians A and B
    - **D.** Neither Technician A nor B

12. *Technician A* says that the generator is responsible for hazardous waste until it reaches the TSDF. *Technician B* says that the generator is responsible for hazardous waste until it is picked up by the waste hauler. Who is correct?
    - **A.** Technician A only
    - **B.** Technician B only
    - **C.** Both Technicians A and B
    - **D.** Neither Technician A nor B

13. *Technician A* says that the manifest is a list of all the hazardous materials used in the shop. *Technician B* says that the manifest is a list of regulations that must be followed by any facility using hazardous materials. Who is correct?
    - **A.** Technician A only
    - **B.** Technician B only
    - **C.** Both Technicians A and B
    - **D.** Neither Technician A nor B

14. *Technician A* says that professional work habits can help create a safer work environment. *Technician B* says that everyone in the shop must play a role in making it a safer workplace. Who is correct?
    - **A.** Technician A only
    - **B.** Technician B only
    - **C.** Both Technicians A and B
    - **D.** Neither Technician A nor B

15. *Technician A* says that a hazardous waste must exhibit four different characteristics to be classified as such. *Technician B* says that it is a hazardous waste if it is considered toxic. Who is correct?
    - **A.** Technician A only
    - **B.** Technician B only
    - **C.** Both Technicians A and B
    - **D.** Neither Technician A nor B

Chapter 3

Hand Tools

■ WORK ASSIGNMENT 3-1

HAMMERS

Name _____   Date _____

Class _____   Instructor _____   Grade _____

1. After reading the assignment, in the space provided below, list the personal and environmental safety equipment and precautions needed for this assignment.

   _____

   _____

   _____

2. Why do you think that hammers, and knowing how to use and maintain them, are so important to a collision repair technician? Record your findings:

   _____

   _____

   _____

3. What is the difference between a ball peen hammer and a dead blow hammer? What is each used for? Record your findings:

   _____

   _____

   _____

Bumping Hammers

4. Why are there so many body hammers? Could a body technician not be successful with only one? Record your findings:

   _____

   _____

   _____

5. Explain the difference between a finishing hammer and a bumping hammer. Record your findings:

   _____

   _____

   _____

6. Why are there so many different sizes of pick hammers? Record your findings:

   _____

   _____

   _____

7. Describe how to repair the face of a hammer. Record your findings:

_____

_____

_____

8. Describe how to install a new hammer handle. Record your findings:

_____

_____

_____

_____

INSTRUCTOR COMMENTS:

_____

_____

_____

_____

_____

_____

_____

DOLLIES AND SPOONS

Name _____ Date _____

Class _____ Instructor _____ Grade _____

1. After reading the assignment, in the space provided below, list the personal and environmental safety equipment and precautions needed for this assignment.

   _____

   _____

   _____

2. What are dollies used for? Record your findings:

   _____

   _____

   _____

3. What is a spoon, and what is it used for? Record your findings:

   _____

   _____

   _____

4. Explain the different uses for general purpose dollies, heel dollies, and toe dollies. Record your findings:

   _____

   _____

   _____

5. How should a dolly be maintained? Record your findings:

   _____

   _____

   _____

6. Because a dolly is often made of hardened steel, why would overheating one during maintenance be dangerous? Explain. Record your findings:

   _____

   _____

   _____

7. What is a body spoon, and what purpose does it serve? Record your findings:

   _____

   _____

   _____

8. What is a spoon dolly and why is it needed? Record your findings:

   _____

   _____

   _____

9. Describe the difference between dinging spoons and surfacing spoons. Record your findings:

   _____

   _____

   _____

10. What is a bumping file or slapping spoon used for? Record your findings:

_____

_____

_____

INSTRUCTOR COMMENTS:

_____

_____

_____

_____

_____

_____

SHAPING AND MISCELLANEOUS TOOLS

Name _____ Date _____

Class _____ Instructor _____ Grade _____

1. After reading the assignment, in the space provided below, list the personal and environmental safety equipment and precautions needed for this assignment.

   _____

   _____

   _____

2. What are caulking irons and what are they used for? Record your findings:

   _____

   _____

3. What are prybars used for, and why are they important in collision work? Record your findings:

   _____

   _____

   _____

4. Describe a weld-on dent pulling tool and how it is used. Record your findings:

   _____

   _____

5. Dent pullers that require a hole are not used as often as they once were. Why? Record your findings:

   _____

   _____

   _____

6. Describe what a Cleco clamp is and how it is used. Record your findings:

   _____

   _____

   _____

7. What is a body file used for? Record your findings:

   _____

   _____

8. What is a board file used for? Record your findings:

   _____

   _____

9. Why does a body technician need assorted blocks? Record your findings:

   _____

   _____

   _____

10. List other specialty tools that a collision technician may need. Record your findings:

_____

_____

_____

INSTRUCTOR COMMENTS:

_____

_____

_____

_____

_____

_____

GENERAL PURPOSE TOOLS

Name _____ Date _____

Class _____ Instructor _____ Grade _____

OBJECTIVES

- Explain/demonstrate the correct and safe use of general purpose hand tools.
- Identify the general purpose hand tools used in collision repair.
- Identify common collision repair hand tools.
- Explain the correct and safe use of collision repair tools.
- Select the correct tools for the job being performed.
- Explain how to properly maintain hand tools.
- Understand how to safely use hand tools.

NATEF TASK CORRELATION

The written and hands-on activities in this chapter satisfy the NATEF High Priority-Individual and High Priority-Group requirements. Though there are no NATEF requirements for this section, it is felt that a working knowledge of the industry is necessary.

Tools and equipment needed (NATEF tool list)

- Pen and pencil
- Safety glasses
- Gloves
- Assorted hand tools

Instructions

The activities for this section are intended to acquaint you with the hand tools. Your instructor will train each of you on the safe use of the tools that you will identify. Safety is a primary concern, and you must follow all instructions, both written and demonstrated, and should not operate equipment or use tools without the expressed direction of your instructor.

PROCEDURE

In the lab, after completing the work assignment for that section, follow the work order and record your findings for the instructor to review. If any section calls for "instructor approval," you should not proceed without this approval.

1. After reading the work order, gather the safety gear needed to complete the task. In the space provided below, list the personal and environmental safety equipment and precautions needed for this assignment. Have the instructor check and approve your plan before proceeding.

_____

_____

_____

INSTRUCTOR'S APPROVAL _____

2. In the work area, find and identify the tools listed below. Then describe the use of each one in the space listed.

Wrenches

_____

_____

_____

½-inch open-end wrench:

_____

_____

_____

13 mm 6-point box-end wrench:

_____

_____

_____

19 mm box-end wrench:

_____

_____

_____

9/16 flare nut wrench:

_____

_____

_____

12-inch adjustable wrench:

_____

_____

_____

13 mm 6-point deep impact socket:

_____

_____

_____

5/16 shallow socket:

_____

_____

_____

#2 Philips head screwdriver:

_____

_____

_____

Needle nose pliers:

_____

_____

_____

Channel lock pliers:

_____

_____

_____

INSTRUCTOR COMMENTS:

_____

_____

_____

_____

_____

_____

HAMMERS

Name _____   Date _____

Class _____   Instructor _____   Grade _____

OBJECTIVES

- Explain/demonstrate the correct and safe use of general purpose hand tools.
- Identify the general purpose hand tools used in collision repair.
- Identify common collision repair hand tools.
- Explain the correct and safe use of collision repair tools.
- Select the correct tools for the job being performed.
- Explain how to properly maintain hand tools.
- Understand how to safely use hand tools.

NATEF TASK CORRELATION

The written and hands-on activities in this chapter satisfy the NATEF High Priority-Individual and High Priority-Group requirements. Though there are no NATEF requirements for this section, it is felt that a working knowledge of the industry is necessary.

Tools and equipment needed (NATEF tool list)

- Pen and pencil
- Safety glasses
- Gloves
- Assorted hand tools

Instructions

The activities for this section are intended to acquaint you with the hand tools. Your instructor will train each of you on the safe use of the tools that you will identify. Safety is a primary concern, and you must follow all instructions, both written and demonstrated, and should not operate equipment or use tools without the expressed direction of your instructor.

PROCEDURE

In the lab, after completing the work assignment for that section, follow the work order and record your findings for the instructor to review. If any section calls for "instructor approval," you should not proceed without this approval.

1. After reading the work order, gather the safety gear needed to complete the task. In the space provided below, list the personal and environmental safety equipment and precautions needed for this assignment. Have the instructor check and approve your plan before proceeding.

_____

_____

_____

INSTRUCTOR'S APPROVAL _____

2. In the work area, find and identify the tools listed below. Then describe the use of each one in the space listed.

   Ball peen hammer:

   _____

   _____

   _____

   Dead blow hammer:

   _____

   _____

   _____

   Bumping hammer:

   _____

   _____

   _____

   Finish hammer:

   _____

   _____

   _____

   2-inch and 4-inch pick hammers:

   _____

   _____

   _____

3. Gather the tools needed to maintain a hammer face. Describe the maintenance procedure below.

   _____

   _____

   _____

INSTRUCTOR COMMENTS:

_____

_____

_____

_____

_____

_____

DOLLIES AND SPOONS

Name _____   Date _____

Class _____   Instructor _____   Grade _____

OBJECTIVES

- Explain/demonstrate the correct and safe use of general purpose hand tools.
- Identify the general purpose hand tools used in collision repair.
- Identify common collision repair hand tools.
- Explain the correct and safe use of collision repair tools.
- Select the correct tools for the job being performed.
- Explain how to properly maintain hand tools.
- Understand how to safely use hand tools.

NATEF TASK CORRELATION

The written and hands-on activities in this chapter satisfy the NATEF High Priority-Individual and High Priority-Group requirements. Though there are no NATEF requirements for this section, it is felt that a working knowledge of the industry is necessary.

Tools and equipment needed (NATEF tool list)

- Pen and pencil
- Safety glasses
- Gloves
- Assorted hand tools

Instructions

The activities for this section are intended to acquaint you with the hand tools. Your instructor will train each of you on the safe use of the tools that you will identify. Safety is a primary concern, and you must follow all instructions, both written and demonstrated, and should not operate equipment or use tools without the expressed direction of your instructor.

PROCEDURE

In the lab, after completing the work assignment for that section, follow the work order and record your findings for the instructor to review. If any section calls for "instructor approval," you should not proceed without this approval.

1. After reading the work order, gather the safety gear needed to complete the task. In the space provided below, list the personal and environmental safety equipment and precautions needed for this assignment. Have the instructor check and approve your plan before proceeding.

_____

_____

_____

INSTRUCTOR'S APPROVAL _____

2. In the work area, find and identify the tools listed below. Then describe the use of each one in the space listed.

   General purpose dolly:

   _____
   _____
   _____

   Heel dolly:

   _____
   _____
   _____

   Toe dolly:

   _____
   _____
   _____

   Dolly spoon:

   _____
   _____
   _____

   Dinging spoon:

   _____
   _____
   _____

3. Gather the tools needed to maintain a spoon face. Describe the maintenance procedure below.

   _____
   _____
   _____

INSTRUCTOR COMMENTS:

_____
_____
_____
_____
_____
_____

SHAPING AND MISCELLANEOUS TOOLS

Name _____   Date _____

Class _____   Instructor _____   Grade _____

OBJECTIVES

- Explain/demonstrate the correct and safe use of general purpose hand tools.
- Identify the general purpose hand tools used in collision repair.
- Identify common collision repair hand tools.
- Explain the correct and safe use of collision repair tools.
- Select the correct tools for the job being performed.
- Explain how to properly maintain hand tools.
- Understand how to safely use hand tools.

NATEF TASK CORRELATION

The written and hands-on activities in this chapter satisfy the NATEF High Priority-Individual and High Priority-Group requirements. Though there are no NATEF requirements for this section, it is felt that a working knowledge of the industry is necessary.

Tools and equipment needed (NATEF tool list)

- Pen and pencil
- Safety glasses
- Gloves
- Assorted hand tools

Instructions

The activities for this section are intended to acquaint you with the hand tools. Your instructor will train each of you on the safe use of the tools that you will identify. Safety is a primary concern, and you must follow all instructions, both written and demonstrated, and should not operate equipment or use tools without the expressed direction of your instructor.

PROCEDURE

In the lab, after completing the work assignment for that section, follow the work order and record your findings for the instructor to review. If any section calls for "instructor approval," you should not proceed without this approval.

1. After reading the work order, gather the safety gear needed to complete the task. In the space provided below, list the personal and environmental safety equipment and precautions needed for this assignment. Have the instructor check and approve your plan before proceeding.

_____

_____

_____

INSTRUCTOR'S APPROVAL _____

2. In the work area, find and identify the tools listed below. Then describe the use of each one in the space listed.

Short and long prybar:

_____

_____

_____

Weld-on pulling gun and studs:

_____

_____

_____

Wiggle wire and pulling handle:

_____

_____

_____

Cleco clamps and tools:

_____

_____

_____

Round and flat body files:

_____

_____

_____

Assorted blocks:

_____

_____

_____

Board file:

_____

_____

_____

INSTRUCTOR COMMENTS:

_____

_____

_____

_____

_____

1. *Technician A* says that an open-end wrench provides the best grip on a hex head fastener. *Technician B* says that a box-end wrench provides the best grip on a hex head fastener. Who is correct?
   A. Technician A only
   B. Technician B only
   C. Both Technicians A and B
   D. Neither Technician A nor B

2. *Technician A* says that a ball peen hammer should be used to drive a chisel. *Technician B* says that a bumping hammer should be used to shape the final contour of a damaged sheet metal panel. Who is correct?
   A. Technician A only
   B. Technician B only
   C. Both Technicians A and B
   D. Neither Technician A nor B

3. *Technician A* says that a flare nut wrench should be used to loosen fuel and brake lines. *Technician B* says that a box-end wrench will provide the best grip for removing fuel and brake lines. Who is correct?
   A. Technician A only
   B. Technician B only
   C. Both Technicians A and B
   D. Neither Technician A nor B

4. *Technician A* says that a Pozidrive screwdriver should be used to remove a clutch head screw. *Technician B* says that a Phillips screwdriver has more grip and less slippage than a Pozidrive screwdriver. Who is correct?
   A. Technician A only
   B. Technician B only
   C. Both Technicians A and B
   D. Neither Technician A nor B

5. *Technician A* says that a Torx driver will have greater turning power and less slippage than a standard screwdriver. *Technician B* says that adjustable pliers are used to remove clips and wires. Who is correct?
   A. Technician A only
   B. Technician B only
   C. Both Technicians A and B
   D. Neither Technician A nor B

6. Which is the best hammer to use for initial straightening on dented panels?
   A. pick hammer
   B. finishing hammer
   C. bumping hammer
   D. dead blow hammer

7. *Technician A* says that a spoon is designed to distribute the striking force over a wide area. *Technician B* says that holes can be drilled to provide access to the damage when using a prybar. Who is correct?
   A. Technician A only
   B. Technician B only
   C. Both Technicians A and B
   D. Neither Technician A nor B

8. *Technician A* says that a standard screwdriver can be used as a chisel. *Technician B* says that a chisel is used to shear off rusted bolt heads. Who is correct?
   A. Technician A only
   B. Technician B only
   C. Both Technicians A and B
   D. Neither Technician A nor B

9. Which of the following should be used to cut external threads?
   A. top
   B. die
   C. Allen wrench
   D. none of the above

10. *Technician A* says that locking pliers will have more holding power than a box-end wrench when loosening an undamaged nut. *Technician B* says that an open-end wrench may slip if too much force is applied. Who is correct?
    A. Technician A only
    B. Technician B only
    C. Both Technicians A and B
    D. Neither Technician A nor B

Chapter 4

Power Tools

■ WORK ASSIGNMENT 4-1

PNEUMATIC AND ELECTRIC POWER

Name _____ Date _____

Class _____ Instructor _____ Grade _____

1. After reading the assignment, in the space provided below, list the personal and environmental safety equipment and precautions needed for this assignment.

   _____

   _____

   _____

2. Describe how pneumatic and electric driven tools differ. Record your findings:

   _____

   _____

   _____

3. What are the inherent dangers when using power tools? Record your findings:

   _____

   _____

   _____

4. Which type of a buffer electric or pneumatic would you use when working in a wet area? Why? Record your findings:

   _____

   _____

   _____

5. What daily tool maintenance do air tools require? Why? Record your findings:

   _____

   _____

6. What is an air ratchet and why would a technician use it? Record your findings:

   _____

   _____

   _____

7. What is torque and how does it apply to a power ratchet? Record your findings:

   _____

   _____

8. Should a technician choose a specific socket design when using a power wrench? Why? Record your findings:

_____

_____

_____

9. How is an air impact ranch different from an air ratchet? Record your findings:

_____

_____

_____

10. What personal safety equipment should a body technician use when operating a grinder? Record your findings:

_____

_____

_____

11. How does a power buffer differ from a power grinder? Record your findings:

_____

_____

_____

12. How should a technician using an electric tool protect from shock? Record your findings:

_____

_____

_____

13. What is a cutoff tool and what safety precautions should a technician use? Record your findings:

_____

_____

_____

14. What is a reciprocating saw and what safety equipment (PPE) should a technician use? Record your findings:

_____

_____

_____

INSTRUCTOR COMMENTS:

_____

_____

_____

_____

_____

_____

BATTERY (CORDLESS) OPERATED AND SANDING TOOLS

Name _____ Date _____

Class _____ Instructor _____ Grade _____

1. After reading the assignment, in the space provided below, list the personal and environmental safety equipment and precautions needed for this assignment.

_____
_____
_____

2. Describe what a cordless tool is and its advantages. Record your findings:

_____
_____
_____

3. What are the inherent dangers when using a cordless drill? Record your findings:

_____
_____
_____

4. Why would a technician choose a cordless ratchet over a pneumatic one? Record your findings:

_____
_____
_____

5. What advantage does a cordless screwdriver have? Record your findings:

_____
_____

6. What advantage does an 18-volt cordless tool have over one with less voltage? Record your findings:

_____
_____
_____

7. What is a straight line sander? Record your findings:

_____
_____
_____

8. How do oscillating and orbital tools differ? Record your findings:

_____
_____
_____

9. How are sanding tools lubricated, and why is this important? Record your findings:

_____
_____
_____

10. Why would a technician need a geared sander? Record your findings:

_____

_____

_____

11. What is a dual action sander used for? Record your findings:

_____

_____

_____

12. What makes a Micro DA different from a standard DA? Record your findings:

_____

_____

_____

13. What is a finish DA, and when would it be used? Record your findings:

_____

_____

_____

INSTRUCTOR COMMENTS:

_____

_____

_____

_____

_____

_____

PNEUMATIC AND ELECTRIC POWER

Name _____   Date _____

Class _____   Instructor _____   Grade _____

OBJECTIVES

- Cite the safe practices that must be employed whenever using power tools.
- Explain/demonstrate the correct and safe use of general purpose hand tools.
- Name and identify pneumatic tools used in collision repair and their use.
- Name and identify electric-powered tools used in collision repair and their use.
- Name and identify battery-powered tools used in collision repair and their use.
- Name and identify hydraulic-powered tools and equipment used in collision repair and their use
- Name and identify welders used in collision repair and their use.
- Name and identify shop equipment used in collision repair and their use.

NATEF TASK CORRELATION

The written and hands-on activities in this chapter satisfy the NATEF High Priority-Individual and High Priority-Group requirements. Though there are no NATEF requirements for this section, it is felt that a working knowledge of the industry is necessary.

Tools and equipment needed (NATEF tool list)

- Pen and pencil
- Safety glasses
- Gloves
- Assorted hand tools

Instructions

The activities for this section are intended to acquaint you with the hand tools. Your instructor will train each of you on the safe use of the tools that you will identify. Safety is a primary concern and you must follow all instructions, both written and demonstrated, and should not operate equipment or use tools without the expressed direction of your instructor.

PROCEDURE

In the lab, after completing the work assignment for that section, follow the work order and record your findings for your instructor to review. If the section calls for "instructor approval," you should not proceed without this approval.

1. After reading the work order, gather the safety gear needed to complete the task. In the space provided below, list the personal and environmental safety equipment and precautions needed for this assignment. Have the instructor check and approve your plan before proceeding.

_____

_____

_____

INSTRUCTOR'S APPROVAL _____

2. In the work area, find and identify the tools listed below, then describe their use in the space listed.

Pneumatic screwdriver # 2 head

_____

_____

_____

What personal protective equipment must a technician use when operating an electric tool?

_____

_____

_____

Demonstrate how to oil a pneumatic tool.

_____

_____

_____

An air ratchet

_____

_____

_____

An electric impact

_____

_____

_____

Find a ½-inch impact socket.

_____

_____

_____

An electric buffer

_____

_____

_____

A pneumatic grinder

_____

_____

_____

A cutoff wheel

_____

_____

_____

A reciprocating saw

_____

_____

_____

INSTRUCTOR COMMENTS:

_____

_____

_____

_____

BATTERY (CORDLESS) OPERATED AND SANDING TOOLS

Name _____   Date _____

Class _____   Instructor _____   Grade _____

OBJECTIVES

- Cite the safe practices that must be employed whenever using power tools.
- Explain/demonstrate the correct and safe use of general purpose hand tools.
- Name and identify pneumatic tools used in collision repair and their use.
- Name and identify electric-powered tools used in collision repair and their use.
- Name and identify battery-powered tools used in collision repair and their use.
- Name and identify hydraulic-powered tools and equipment used in collision repair and their use.
- Name and identify welders used in collision repair and their use.
- Name and identify shop equipment used in collision repair and their use.

NATEF TASK CORRELATION

The written and hands-on activities in this chapter satisfy the NATEF High Priority-Individual and High Priority-Group requirements. Though there are no NATEF requirements for this section, it is felt that a working knowledge of the industry is necessary.

Tools and equipment needed (NATEF tool list)

- Pen and pencil
- Safety glasses
- Gloves
- Assorted hand tools

Instructions

The activities for this section are intended to acquaint you with the hand tools. Your instructor will train each of you on the safe use of the tools that you will identify. Safety is a primary concern and you must follow all instructions, both written and demonstrated, and should not operate equipment or use tools without the expressed direction of your instructor.

PROCEDURE

In the lab, after completing the work assignment for that section, follow the work order and record your findings for your instructor to review. If the section calls for "instructor approval," you should not proceed without this approval.

1. After reading the work order, gather the safety gear needed to complete the task. In the space provided below, list the personal and environmental safety equipment and precautions needed for this assignment. Have the instructor check and approve your plan before proceeding.

_____

_____

_____

INSTRUCTOR'S APPROVAL _____

2. In the work area find and identify the tools listed below, then describe their use in the space listed.

Cordless trouble light:

_____

_____

Cordless screwdriver:

_____

_____

Charging station:

_____

_____

Straight line sander:

_____

_____

An oscillating sander:

_____

_____

An orbital sander:

_____

_____

A grinder:

_____

_____

A DA:

_____

_____

A finish DA:

_____

_____

_____

INSTRUCTOR COMMENTS:

_____

_____

_____

_____

_____

WHEELED, BENCH-MOUNTED, AND HYDRAULIC TOOLS

Name _____ Date _____

Class _____ Instructor _____ Grade _____

OBJECTIVES

- Cite the safe practices that must be employed whenever using power tools.
- Explain/demonstrate the correct and safe use of general purpose hand tools.
- Name and identify pneumatic tools used in collision repair and their use.
- Name and identify electric-powered tools used in collision repair and their use.
- Name and identify battery-powered tools used in collision repair and their use.
- Name and identify hydraulic-powered tools and equipment used in collision repair and their use.
- Name and identify welders used in collision repair and their use.
- Name and identify shop equipment used in collision repair and their use.

NATEF TASK CORRELATION

The written and hands-on activities in this chapter satisfy the NATEF High Priority-Individual and High Priority-Group requirements. Though there are no NATEF requirements for this section, it is felt that a working knowledge of the industry is necessary.

Tools and equipment needed (NATEF tool list)

- Pen and pencil
- Safety glasses
- Gloves
- Assorted hand tools

Instructions

The activities for this section are intended to acquaint you with the hand tools. Your instructor will train each of you on the safe use of the tools that you will identify. Safety is a primary concern and you must follow all instructions, both written and demonstrated, and should not operate equipment or use tools without the expressed direction of your instructor.

PROCEDURE

In the lab, after completing the work assignment for that section, follow the work order and record your findings for your instructor to review. If the section calls for "instructor approval," you should not proceed without this approval.

1. After reading the work order, gather the safety gear needed to complete the task. In the space provided below, list the personal and environmental safety equipment and precautions needed for this assignment. Have the instructor check and approve your plan before proceeding.

_____

_____

_____

INSTRUCTOR'S APPROVAL _____

2. In the work area find and identify the tools listed below, then describe their use in the space listed.
A 110v GMAW (MIG) welder:

_____

_____

_____

A squeeze-type welder:

_____

_____

_____

Induction tool:

_____

_____

_____

Bench grinder:

_____

_____

_____

Drill press:

_____

_____

_____

Port-a-Power:

_____

_____

_____

Frame bench:

_____

_____

_____

INSTRUCTOR COMMENTS:

_____

_____

_____

_____

_____

_____

Name _____   Date _____

Class _____   Instructor _____   Grade _____

1. *Technician A* says that electric tools are the technicians' tool of choice because they are lighter weight than air tools. *Technician B* says that a right angle grinder that has a disc speed of 2,500 rpm is the preferred choice for paint and filler removal. Who is correct?
   A. Technician A only
   B. Technician B only
   C. Both Technicians A and B
   D. Neither Technician A nor B

2. *Technician A* says that chrome-plated hand sockets can safely be used with an air ratchet. *Technician B* says that the power of an air ratchet is determined by the number of veins on the drive rotor. Who is correct?
   A. Technician A only
   B. Technician B only
   C. Both Technicians A and B
   D. Neither Technician A nor B

3. *Technician A* says that the capacity of a drill is determined by the largest size arbor that the chuck is able to hold. *Technician B* says that the nibbler is used to break spot welds for panel removal. Who is correct?
   A. Technician A only
   B. Technician B only
   C. Both Technicians A and B
   D. Neither Technician A nor B

4. *Technician A* says that whenever using an air ratchet, the trigger should be depressed until the tool stalls out and then finish tightening the fastener by hand. *Technician B* says that the tool rest on the bench grinder should be adjusted to within 1/8 inch of the grinding wheel. Who is correct?
   A. Technician A only
   B. Technician B only
   C. Both Technicians A and B
   D. Neither Technician A nor B

5. *Technician A* says that the air hammer or chisel is frequently used for removing corroded bolts and nuts. *Technician B* says that the die grinder is commonly used to dress damaged and corroded bolt threads. Who is correct?
   A. Technician A only
   B. Technician B only
   C. Both Technicians A and B
   D. Neither Technician A nor B

6. *Technician A* says that using a GFCI will help an electric tool to run cooler. *Technician B* says that the purpose for using a GFCI is to help avoid electric shock. Who is correct?
   A. Technician A only
   B. Technician B only
   C. Both Technicians A and B
   D. Neither Technician A nor B

7. *Technician A* says that an impact wrench continues to tighten the fastener by a hammering effect after it has stalled out. *Technician B* says that an air ratchet turns the bolt or nut with a continuous driving motion until it stalls out and then it stops. Who is correct?
   A. Technician A only
   B. Technician B only
   C. Both Technicians A and B
   D. Neither Technician A nor B

8. *Technician A* says that the heavy-duty right-angle grinder is the best choice for removing excess weld material that may occur during a repair process. *Technician B* says that most pistol grip grinders rotate at a rate of approximately 10,000 rpm. Who is correct?
   A. Technician A only
   B. Technician B only
   C. Both Technicians A and B
   D. Neither Technician A nor B

9. *Technician A* says that the induction heater is a flameless operation, making it ideal to use around gas lines, fuel cells, and the like. *Technician B* says that the impact gun may be used for installing wheel lugnuts as long as they are used with impact socket sticks. Who is correct?
   A. Technician A only
   B. Technician B only
   C. Both Technicians A and B
   D. Neither Technician A nor B

10. *Technician A* says that an air tool should be lubricated with at least six drops of oil before putting it away after each time it is used. *Technician B* says that the cutoff saw blade should be touching the material being cut before it is turned on. Who is correct?
    A. Technician A only
    B. Technician B only
    C. Both Technicians A and B
    D. Neither Technician A nor B

Chapter 5

Welding Procedures & Equipment

■ WORK ASSIGNMENT 5-1

SET UP ADJUST GMAW (MIG) WELDS

Name _____   Date _____

Class _____   Instructor _____   Grade _____

NATEF TASKS E. 1, 2, 6, 7, 11

1. After reading the assignment, in the space provided below, list the personal and environmental safety equipment and precautions needed for this assignment.

   _____

   _____

   _____

2. Explain how a collision repair technician would determine if an automotive part is weldable or not.

   _____

   _____

   _____

3. Using the welder's operator manual and industry setup recommendations, list how a MIG welder would be set up for automotive welding.
   Welder size and type _____

   Wire size used _____

   Electrode type _____

   Wire type _____

   Electrode diameter_____

   Shielding gas _____

4. Using 18-gauge welding coupons, "tune" the welder.
   Electrode stakeout _____

   Voltage _____

   Polarity used _____

   Gas flow rate _____

   Wire speed used_____

5. Explain the proper installation, handling, and storage procedures for high-pressure gas.

_____

_____

_____

6. Explain the importance of the welder's work clamp.

_____

_____

_____

7. Explain the importance of the location of the work clamp relative to electronic equipment.

_____

_____

_____

8. Describe how to remove, safely store, and reinstall the high-pressure gas cylinders.

_____

_____

_____

INSTRUCTOR COMMENTS:

_____

_____

_____

_____

_____

_____

PRACTICE WELDS

Name _____   Date _____

Class _____   Instructor _____   Grade _____

NATEF II NONSTRUCTURAL AND DAMAGE REPAIR TASKS E. 1, 2, 3, 4, 5, 7, 8, 11, 12, 13, 14, 15, 16

After reading the assignment, in the space provided below, list the personal and environmental safety equipment and precautions needed for this assignment.

1. Set up the welder for welding 18-gauge steel.
   Welder size and type _____

   Wire size used _____

   Electrode type _____

   Wire type _____

   Electrode diameter _____

   Shielding gas _____

2. Explain the proper cleaning and setup for continuous welding of 18-gauge steel. Record your findings:

   _____

   _____

   _____

3. Explain the proper cleaning and setup for aluminum. Record your findings:

   _____

   _____

   _____

4. Explain the proper cleaning and setup for stitch welding 18-gauge steel. Record your findings:

   _____

   _____

   _____

5. Explain the proper cleaning and setup for tack welding 18-gauge steel. Record your findings:

   _____

   _____

   _____

6. Explain the proper cleaning and setup for plug welding 18-gauge steel. Record your findings:

   _____

   _____

   _____

7. Explain the proper cleaning and setup for butt with a backer welding 18-gauge steel. Record your findings:

   _____

   _____

   _____

8. Explain the proper cleaning and setup for butt without a backer welding 18-gauge steel. Record your findings:

_____

_____

_____

9. Explain the proper cleaning and setup for lap welding 18-gauge steel. Record your findings:

_____

_____

_____

INSTRUCTOR COMMENTS:

_____

_____

_____

_____

_____

_____

LIVE WELDING

Name _____ Date _____

Class _____ Instructor _____ Grade _____

NATEF TASKS E. 9, 10, 13, 17, 18, 19

After reading the assignment, in the space provided below, list the personal and environmental safety equipment and precautions needed for this assignment.

1. Set up the welder for welding 18-gauge steel.
    Welder size and type_____

    Wire size used _____

    Electrode type _____

    Wire type_____

    Electrode diameter _____

    Shielding gas _____

2. Explain the proper cleaning and setup welding on an automatable. Record your findings:

    _____

    _____

    _____

3. An S10 truck needs its radiator support replaced. How would you protect the glass of this vehicle? Record your findings:

    _____

    _____

    _____

4. How would you correctly protect the electronic control modules on this vehicle? Record your findings:

    _____

    _____

    _____

5. How would you find the recommended weld method and type to attach a core support on a Toyota pickup? Record your findings:

    _____

    _____

    _____

6. While welding the welder has tip burn back and stops welding. What is causing this condition and how would you correct it? Record your findings:

    _____

    _____

    _____

7. Describe the different types of cutting methods that can be used in collision repair for both steel and aluminum. Record your findings:

_____

_____

_____

8. Describe the different methods for attaching nonstructural components.

_____

_____

_____

INSTRUCTOR COMMENTS:

_____

_____

_____

_____

_____

_____

SET UP ADJUST GMAW (MIG) WELDS

Name _____   Date _____

Class _____   Instructor _____   Grade _____

OBJECTIVES

- Know, understand, and use the safety equipment necessary for the task.
- Show proficiency in setup, tuning, and operation of a welder for the task at hand.
- Perform acceptable lap welds on steel.
- Perform acceptable plug welds on aluminum.
- Perform acceptable continuous welds on steel.
- Perform acceptable butt welds without a backer on steel.
- Perform acceptable butt welds with a backer on steel.
- Recognize and repair causes for no feed of a MIG welder as necessary.

NATEF TASK CORRELATION

The written and hands-on activities in this chapter satisfy the NATEF High Priority-Individual and High Priority-Group requirements **for** Section II: Nonstructural and Damage Repair Tasks E. 1, 2, 3, 6, 7, 11.

Tools and equipment needed (NATEF tool list)

- Pencil and paper
- Safety glasses
- Gloves
- Ear protection
- Particle mask
- Steel welding equipment
- Aluminum welding equipment

PROCEDURE

1. After reading the assignment, in the space provided below, list the personal and environmental safety equipment and precautions needed for this assignment

_____

_____

_____

INSTRUCTOR'S APPROVAL _____

2. Using the items provided, determine if they are weldable or nonweldable. List the items:

_____

_____

3. Describe the safety precautions that should be observed to safely change and store a high-pressure gas cylinder.

_____

_____

_____

4. After instructor approval, remove and safely store a high-pressure gas cylinder, then safely reinstall a high-pressure gas cylinder.

5. Load the wire, attach the gas cylinder, and adjust the machine for welding 18-gauge steel.
   Welder size and type _____

   Wire size used _____

   Electrode type _____

   Wire type _____

   Electrode diameter _____

   Shielding gas _____

   Have the instructor evaluate:

   _____

   _____

   _____

6. Tune the welder and record the setting.
   Electrode stakeout _____

   Voltage _____

   Polarity used _____

   Gas flow rate _____

   Wire speed used _____

   Have the instructor evaluate:

   _____

   _____

   _____

7. Attach the work clamp, clean and prepare the steel for welding. Then perform four acceptable 4-inch lap welds. Have the instructor evaluate:
   Weld 1

   _____

   _____

   _____

   Weld 2

   _____

   _____

   _____

   Weld 3

   _____

   _____

   _____

   Weld 4

   _____

   _____

   _____

8. Attach the work clamp, clean and prepare the aluminum for welding. Then perform four acceptable 4-inch lap welds. Have the instructor evaluate:

Weld 1

_____
_____
_____

Weld 2

_____
_____
_____

Weld 3

_____
_____
_____

Weld 4

_____
_____
_____

INSTRUCTOR COMMENTS:

_____
_____
_____
_____
_____
_____

PRACTICE WELDS

Name _____   Date _____

Class _____   Instructor _____   Grade _____

OBJECTIVES

- Know, understand, and use the safety equipment necessary for the task.
- Show proficiency in setup, tuning, and operation of a welder for the task at hand.
- Perform acceptable continuous welds on steel.
- Perform acceptable continuous welds on aluminum.
- Perform acceptable stitch welds on steel.
- Perform acceptable tack welds on steel.
- Recognize and repair causes for no feed of a MIG welder as necessary.

NATEF TASK CORRELATION

The written and hands-on activities in this chapter satisfy the NATEF High Priority-Individual and High Priority-Group requirements for Section II: Nonstructural and Damage Repair Tasks E. 1, 2, 3, 4, 5, 7, 8, 11, 12, 13, 14, 15, 16.

Tools and equipment needed (NATEF tool list)

- Pencil and paper
- Safety glasses
- Gloves
- Ear protection
- Partial mask
- Steel welding equipment
- Aluminum welding equipment

PROCEDURE

1. After reading the work order, gather the safety gear needed to complete the task. In the space provided below, list the personal and environmental safety equipment and precautions needed for this assignment. Have the instructor check and approve your plan before proceeding.

_____

_____

_____

INSTRUCTOR'S APPROVAL _____

2. Set up the welder for welding 18-gauge steel.
   Welder size and type_____

   Wire size used _____

   Electrode type _____

   Wire type_____

   Electrode diameter _____

   Shielding gas _____

Have the instructor evaluate:

_____

_____

_____

3. Tune the welder and record the setting.
   Electrode stakeout_____

   Voltage _____

   Polarity used _____

   Gas flow rate _____

   Wire speed used _____

   Have the instructor evaluate:

   _____

   _____

   _____

4. Attach the work clamp, clean and prepare the metal for welding, then perform four 4-inch acceptable continuous welds. Have the instructor evaluate:
   Weld 1

   _____

   _____

   Weld 2

   _____

   _____

   Weld 3

   _____

   _____

   Weld 4

   _____

   _____

5. Attach the work clamp, clean and prepare the aluminum for welding, then perform four 4-inch acceptable continuous welds. Have the instructor evaluate:
   Weld 1

   _____

   _____

   Weld 2

   _____

   _____

   Weld 3

   _____

   _____

Weld 4

_____

_____

_____

6. Attach the work clamp, clean and prepare the metal for welding, then perform four 4-inch acceptable stitch welds. Have the instructor evaluate:

Weld 1

_____

_____

_____

Weld 2

_____

_____

_____

Weld 3

_____

_____

_____

Weld 4

_____

_____

_____

7. Attach the work clamp, clean and prepare the metal for welding, then perform 20 acceptable tack welds. Have the instructor evaluate:

Weld 1

_____

_____

_____

Weld 2

_____

_____

_____

Weld 3

_____

_____

_____

Weld 4

_____

_____

_____

INSTRUCTOR COMMENTS:

_____

_____

_____

_____

PRACTICE WELDS

Name _____   Date _____

Class _____   Instructor _____   Grade _____

OBJECTIVES

- Know, understand, and use the safety equipment necessary for the task.
- Show proficiency in setup, tuning, and operation of a welder for the task at hand.
- Perform acceptable plug welds on steel.
- Perform acceptable plug welds on aluminum.
- Perform acceptable continuous welds on steel.
- Perform acceptable butt welds without a backer on steel.
- Perform acceptable butt welds with a backer on steel.
- Recognize and repair causes for no feed of a MIG welder as necessary.

NATEF TASK CORRELATION

The written and hands-on activities in this chapter satisfy the NATEF High Priority-Individual and High Priority-Group requirements for Section II: Nonstructural and Damage Repair Tasks E. 1, 2, 3, 4, 5, 7, 8, 11, 12, 13, 14, 15, 16.

Tools and equipment needed (NATEF tool list)

- Pencil and paper
- Safety glasses
- Gloves
- Ear protection
- Partial mask
- Steel welding equipment
- Aluminum welding equipment

PROCEDURE

1. After reading the work order, gather the safety gear needed to complete the task. In the space provided below, list the personal and environmental safety equipment and precautions needed for this assignment. Have the instructor check and approve your plan before proceeding.

_____

_____

_____

INSTRUCTOR'S APPROVAL _____

2. Set up the welder for welding 18-gauge steel.
   Welder size and type _____

   Wire size used _____

   Electrode type _____

   Wire type _____

   Electrode diameter _____

   Shielding gas _____

Have the instructor evaluate:

_____

_____

_____

3. Tune the welder and record the setting.
   Electrode stakeout_____

   Voltage _____

   Polarity used _____

   Gas flow rate _____

   Wire speed used _____

   Have the instructor evaluate:

   _____

   _____

   _____

4. Attach the work clamp, clean and prepare the metal for welding, then perform four 4-inch acceptable plug welds Have the instructor evaluate:
   Weld 1

   _____

   _____

   Weld 2

   _____

   _____

   Weld 3

   _____

   _____

   Weld 4

   _____

   _____

5. Attach the work clamp, clean and prepare the aluminum for welding, then perform 20 acceptable plug welds. Have the instructor evaluate:
   Weld 1

   _____

   _____

   Weld 2

   _____

   _____

   Weld 3

   _____

   _____

Weld 4

_____

_____

_____

6. Attach the work clamp, clean and prepare the metal for welding, then perform four 4-inch acceptable butt with a backer welds. Have the instructor evaluate:

Weld 1

_____

_____

Weld 2

_____

_____

Weld 3

_____

_____

Weld 4

_____

_____

7. Attach the work clamp, clean and prepare the metal for welding, then perform four 4-inch acceptable butt without a backer welds. Have the instructor evaluate:

Weld 1

_____

_____

Weld 2

_____

_____

Weld 3

_____

_____

Weld 4

_____

_____

8. On a welder provided, identify the cause of tip burn-back and failure to feed and then make the necessary adjustments. Describe your findings.

_____

_____

_____

INSTRUCTOR COMMENTS:

_____
_____
_____
_____
_____
_____

Name _____ Date _____

Class _____ Instructor _____ Grade _____

1. *Technician A* says that argon gas is used for shielding MIG welding of automotive steel. *Technician B* says that carbon monoxide gas is used for shielding MIG welding of automotive steel. Who is correct?
    A. Technician A only
    B. Technician B only
    C. Both Technicians A and B
    D. Neither Technician A nor B

2. *Technician A* says that electrode wire 0.030 is used for automotive body steel. *Technician B* says that electrode wire 0.023 is used for automotive body steel. Who is correct?
    A. Technician A only
    B. Technician B only
    C. Both Technicians A and B
    D. Neither Technician A nor B

3. *Technician A* says that safety glasses are not needed under your welding helmet. *Technician B* says that safety glasses should be worn under a welding helmet. Who is correct?
    A. Technician A only
    B. Technician B only
    C. Both Technicians A and B
    D. Neither Technician A nor B

4. *Technician A* says that welding sparks and slag will pit automotive glass. *Technician B* says that welding sparks and slag may even harm the glass of the vehicle close by the one being worked on and it should be protected as well. Who is correct?
    A. Technician A only
    B. Technician B only
    C. Both Technicians A and B
    D. Neither Technician A nor B

5. *Technician A* says that the work clamp should be placed close to the area being welded. *Technician B* says that welding will not damage electronic equipment. Who is correct?
    A. Technician A only
    B. Technician B only
    C. Both Technicians A and B
    D. Neither Technician A nor B

6. *Technician A* says that cleaning and fit-up of the weld site is critical for a good weld. *Technician B* says that the light from welding can cause exposed skin to burn. Who is correct?
    A. Technician A only
    B. Technician B only
    C. Both Technicians A and B
    D. Neither Technician A nor B

7. *Technician A* says that some vehicle manufacturers recommend that a butt weld be made without a backer. *Technician B* says that some vehicle manufacturers recommend that a butt weld be made with a backer. Who is correct?
    A. Technician A only
    B. Technician B only
    C. Both Technicians A and B
    D. Neither Technician A nor B

8. **TRUE** or **FALSE**. All manufacturers recommend the use of weld-through primer when welding.

9. *Technician A* says that while working in an enclosed space, a respirator may be needed because of the smoke produced by welding. *Technician B* says that flammable vehicle parts are removed from the weld area to prevent fires. Who is correct?
    A. Technician A only
    B. Technician B only
    C. Both Technicians A and B
    D. Neither Technician A nor B

10. **TRUE** or **FALSE**. A fire extinguisher should always be close at hand when welding.

Chapter 6

Vehicle Construction

■ WORK ASSIGNMENT 6-1

MATERIALS AND SHAPES USED

Name _____   Date _____

Class _____   Instructor _____   Grade _____

1. After reading the assignment, in the space provided below, list the personal and environmental safety equipment and precautions needed for this assignment.

   _____

   _____

   _____

2. What are the NHTSA, IIHS, and EPA, and how do they affect automotive design and manufacturing? Record your findings:

   _____

   _____

   _____

3. What are the HSS, HSLA, and UHSS? Describe their effects on automotive design. Record your findings:

   _____

   _____

   _____

4. Explain work hardening and how it is used when designing and building an automobile. Record your findings:

   _____

   _____

   _____

5. How do work hardening and the common U shape strengthen a vehicle? Record your findings:

   _____

   _____

   _____

6. How can collision energy be directed to safeguard passengers? Record your findings:

   _____

   _____

   _____

7. Why had aluminum parts on vehicles become more predominant in automotive design?   Record your findings:

_____

_____

_____

8. What part does plastic play in automotive design? Record your findings:

_____

_____

_____

INSTRUCTOR COMMENTS:

_____

_____

_____

_____

_____

_____

BODY OVER FRAME, UNIBODY, AND SPACE FRAME

Name _____ Date _____

Class _____ Instructor _____ Grade _____

1. After reading the assignment, in the space provided below, list the personal and environmental safety equipment and precautions needed for this assignment.

   _____

   _____

   _____

2. Describe body over frame design. Record your findings:

   _____

   _____

   _____

3. Describe a ladder frame and a perimeter frame.

   Ladder:

   _____

   _____

   _____

   Perimeter:

   _____

   _____

   _____

4. What is the significance of a torque box? Record your findings:

   _____

   _____

   _____

5. How does the drop canter design help a vehicle become safer? Record your findings:

   _____

   _____

   _____

6. What is hydroforming? Record your findings:

   _____

   _____

   _____

7. Explain unibody design and its advantages. Record your findings:

   _____

   _____

   _____

8. What is a space frame? Record your findings:

   _____

   _____

   _____

9. What is a subframe and what is it used for? Record your findings:

_____

_____

_____

INSTRUCTOR COMMENTS:

_____

_____

_____

_____

_____

_____

COLLISION MANAGEMENT

Name _____   Date _____

Class _____   Instructor _____   Grade _____

1. After reading the assignment, in the space provided below, list the personal and environmental safety equipment and precautions needed for this assignment.

   _____

   _____

   _____

2. How are collision forces affected by vehicle design? Record your findings:

   _____

   _____

   _____

3. What purpose do the front rails and core support perform in a collision? Record your findings:

   _____

   _____

   _____

4. What is a tailor-welded blank? Record your findings:

   _____

   _____

   _____

5. How does the passenger compartment cage protect its occupants? Record your findings:

   _____

   _____

   _____

6. How do nonstructural parts such as a door protect passengers? Record your findings:

   _____

   _____

   _____

7. How do hoods and deck lids deflect collision energy? Record your findings:

   _____

   _____

   _____

INSTRUCTOR COMMENTS:

_____

_____

_____

_____

_____

_____

MATERIALS AND SHAPES USED

Name _____   Date _____

Class _____   Instructor _____   Grade _____

OBJECTIVES

- Distinguish the difference between the body over frame and unibody designs.
- Identify the role of the frame on a full frame vehicle.
- Identify the different frame designs utilized on the modern day vehicle and the advantages of each.
- Identify the structural design and role of the undercarriage on a unibody design.
- Identify the difference between a space frame design vehicle and a true unibody design vehicle.
- Identify the difference in the attachment of the suspension and steering parts on a unibody vehicle and a body over frame vehicle.
- Identify the difference in the design characteristics of a front-wheel-drive vehicle and a rear-wheel-drive vehicle.
- Identify the differences among the front-engine, mid-engine, and rear-engine designs.
- Discuss the concepts of collision energy management to safeguard the vehicle occupants.
- Review the emerging trends and materials used in vehicle construction.
- Be able to answer all ASE-style, essay, and other critical thinking questions and activities.

NATEF TASK CORRELATION

The written and hands-on activities in this chapter satisfy the NATEF High Priority-Individual and High Priority-Group requirements. Though there are no NATEF requirements for this section, it is felt that a working knowledge of the industry is necessary.

Tools and equipment needed (NATEF tool list)

- Pen and pencil
- Safety glasses
- Gloves
- Pick hammer
- Dolly

Instructions

The activities for this section are intended to acquaint you with the shop and its equipment. Your instructor will train each of you on the safe use of the equipment that you will identify today. Safety is a primary concern and you must follow all instructions, both written and demonstrated, and should not operate equipment without the expressed direction of your instructor.

PROCEDURE

In the lab, after completing the work assignment for that section, follow the work order and record your findings for your instructor to review. If the section calls for "instructor approval," you should not proceed without this approval.

1. After reading the work order, gather the safety gear needed to complete the task. In the space provided below, list the personal and environmental safety equipment and precautions needed for this assignment. Have the instructor check and approve your plan before proceeding.

   _____

   _____

   _____

INSTRUCTOR'S APPROVAL _____

2. Using the vehicle service manuals, identify the high-strength steel on the vehicle in the lab. Record your findings:

   _____

   _____

   _____

3. Using the steel part provided by your instructor, identify and straighten the work-hardened buckle. How did you overcome the work hardening?

   _____

   _____

   _____

4. Using the aluminum part provided by your instructor, identify and straighten the work-hardened buckle. Did it differ from steel?

   _____

   _____

   _____

INSTRUCTOR COMMENTS:

_____

_____

_____

_____

_____

_____

BODY OVER FRAME, UNIBODY, AND SPACE FRAME

Name _____ Date _____

Class _____ Instructor _____ Grade _____

OBJECTIVES

- Distinguish the difference between the body over frame and unibody designs.
- Identify the role of the frame on a full frame vehicle.
- Identify the different frame designs utilized on the modern day vehicle and the advantages of each.
- Identify the structural design and role of the undercarriage on a unibody design.
- Identify the difference between a space frame design vehicle and a true unibody design vehicle.
- Identify the difference in the attachment of the suspension and steering parts on a unibody vehicle and a body over frame vehicle.
- Identify the difference in the design characteristics of a front-wheel-drive vehicle and a rear-wheel-drive vehicle.
- Identify the differences among the front-engine, mid-engine, and rear-engine designs.
- Discuss the concepts of collision energy management to safeguard the vehicle occupants.
- Review the emerging trends and materials used in vehicle construction.
- Be able to answer all ASE-style, essay, and other critical thinking questions and activities.

NATEF TASK CORRELATION

The written and hands-on activities in this chapter satisfy the NATEF High Priority-Individual and High Priority-Group requirements. Though there are no NATEF requirements for this section, it is felt that a working knowledge of the industry is necessary.

Tools and equipment needed (NATEF tool list)

- Pen and pencil
- Safety glasses
- Gloves

Instructions

The activities for this section are intended to acquaint you with the shop and its equipment. Your instructor will train each of you on the safe use of the equipment that you will identify today. Safety is a primary concern and you must follow all instructions, both written and demonstrated, and should not operate equipment without the expressed direction of your instructor.

PROCEDURE

In the lab, after completing the work assignment for that section, follow the work order and record your findings for your instructor to review. If the section calls for "instructor approval," you should not proceed without this approval.

1. After reading the work order, gather the safety gear needed to complete the task. In the space provided below, list the personal and environmental safety equipment and precautions needed for this assignment. Have the instructor check and approve your plan before proceeding.

_____

_____

_____

INSTRUCTOR'S APPROVAL _____

2. Using the vehicles provided in the lab, identify the body and frame design types listed below:

   The body over frame vehicle

   _____

   What design element verifies your choice?

   _____

   _____

   _____

   The ladder frame vehicle:

   _____

   What design element verifies your choice?

   _____

   _____

   _____

   The perimeter frame vehicle:

   _____

   What design element verifies your choice?

   _____

   _____

   _____

3. Where is the torque box? Record your findings:

   _____

   _____

   _____

4. Which vehicle is a unibody design vehicle?

   _____

5. What design element verifies your choice?

   _____

   _____

   _____

6. Which vehicle uses a subframe?

   _____

7. What design element verifies your choice?

   _____

   _____

   _____

INSTRUCTOR COMMENTS:

_____

_____

_____

_____

_____

_____

COLLISION MANAGEMENT

Name _____   Date _____

Class _____   Instructor _____   Grade _____

OBJECTIVES

- Distinguish the difference between the body over frame and unibody designs.
- Identify the role of the frame on a full frame vehicle.
- Identify the different frame designs utilized on the modern day vehicle and the advantages of each.
- Identify the structural design and role of the undercarriage on a unibody design.
- Identify the difference between a space frame design vehicle and a true unibody design vehicle.
- Identify the difference in the attachment of the suspension and steering parts on a unibody vehicle and a body over frame vehicle.
- Identify the difference in the design characteristics of a front-wheel-drive vehicle and a rear-wheel-drive vehicle.
- Identify the differences among the front-engine, mid-engine, and rear-engine designs.
- Discuss the concepts of collision energy management to safeguard the vehicle occupants.
- Review the emerging trends and materials used in vehicle construction.
- Be able to answer all ASE-style, essay, and other critical thinking questions and activities.

NATEF TASK CORRELATION

The written and hands-on activities in this chapter satisfy the NATEF High Priority-Individual and High Priority-Group requirements. Though there are no NATEF requirements for this section, it is felt that a working knowledge of the industry is necessary.

Tools and equipment needed (NATEF tool list)

- Pen and pencil
- Safety glasses
- Gloves

Instructions

The activities for this section are intended to acquaint you with the shop and its equipment. Your instructor will train each of you on the safe use of the equipment that you will identify today. Safety is a primary concern and you must follow all instructions, both written and demonstrated, and should not operate equipment without the expressed direction of your instructor.

PROCEDURE

In the lab, after completing the work assignment for that section, follow the work order and record your findings for your instructor to review. If the section calls for "instructor approval," you should not proceed without this approval.

1. After reading the work order, gather the safety gear needed to complete the task. In the space provided below, list the personal and environmental safety equipment and precautions needed for this assignment. Have the instructor check and approve your plan before proceeding.

_____

_____

_____

INSTRUCTOR'S APPROVAL _____

2. Using the damaged vehicles in the lab, identify and describe the energy and management designs effect.

Energy transmission through the bumper/fasciae:

_____

_____

_____

Energy transmission through the front rails:

_____

_____

_____

Energy transmission through the core support:

_____

_____

_____

Energy transmission through the passenger compartment:

_____

_____

_____

Energy transmission through the hood:

_____

_____

_____

Energy transmission through the floor pan:

_____

_____

_____

Energy transmission through the deck lid:

_____

_____

_____

INSTRUCTOR COMMENTS:

_____

_____

_____

_____

_____

_____

Name _____   Date _____

Class _____   Instructor _____   Grade _____

1. *Technician A* says that the body over frame vehicle was the most popular design in the United States for both cars and trucks prior to the introduction of the unibody. *Technician B* says that the vehicle weight of a unibody is heavier than that of the body-over-frame. Who is correct?
   A. Technician A only
   B. Technician B only
   C. Both Technicians A and B
   D. Neither Technician A nor B

2. *Technician A* says that much of the technology for automobile safety design concepts comes from the aircraft industry. *Technician B* says that the effects of the collision energy travel throughout the vehicle when it is struck. Who is correct?
   A. Technician A only
   B. Technician B only
   C. Both Technicians A and B
   D. Neither Technician A nor B

3. *Technician A* says that all unibody vehicles utilize a stub frame under the front of the vehicle to attach the drivetrain. *Technician B* says that the enclosed box-type construction is usually used throughout the length of the frame rails. Who is correct?
   A. Technician A only
   B. Technician B only
   C. Both Technicians A and B
   D. Neither Technician A nor B

4. *Technician A* says that the manufacturers use plastic sandwiching as a means to enhance the strength of floor pan. *Technician B* says that wood is also used for plastic sandwiching. Who is correct?
   A. Technician A only
   B. Technician B only
   C. Both Technicians A and B
   D. Neither Technician A nor B

5. *Technician A* says that the weight distribution of a unibody is approximately 60% at the front and 40% at the rear of the vehicle. *Technician B* says that the manufacturers use foam to strengthen and reinforce some pillars. Who is correct?
   A. Technician A only
   B. Technician B only
   C. Both Technicians A and B
   D. Neither Technician A nor B

6. In a front-wheel-drive vehicle the engine and transaxle are mounted:
   A. longitudinally
   B. transversely
   C. laterally
   D. in line

7. *Technician A* says that the suspension parts for a unibody with a subframe are mounted to the body. *Technician B* says that a hat channel is commonly used on the rear floor pan for reinforcement. Who is correct?
   A. Technician A only
   B. Technician B only
   C. Both Technicians A and B
   D. Neither Technician A nor B

8. *Technician A* says that the ladder frame is best suited for larger and heavier vehicles. *Technician B* says that the perimeter frame is most often used for larger vehicles. Who is correct?
   A. Technician A only
   B. Technician B only
   C. Both Technicians A and B
   D. Neither Technician A nor B

9. *Technician A* says that stress risers are used to cause the metal to bend at a predetermined location. *Technician B* says that an enclosed boxlike structure is used in areas where maximum strength is required. Who is correct?
   A. Technician A only
   B. Technician B only
   C. Both Technicians A and B
   D. Neither Technician A nor B

10. Which of the following is NOT used to create a crush zone?
    A. increasing the distance between spot welds
    B. piercing slots or holes into the metal
    C. using heavier gauge metal at the front of a rail section
    D. using tailor-welded blanks

11. *Technician A* says that a full frame vehicle is usually a rear-wheel drive. *Technician B* says that the aerodynamics of a vehicle will not affect how fuel efficient it is. Who is correct?
    A. Technician A only
    B. Technician B only
    C. Both Technicians A and B
    D. Neither Technician A nor B

12. Which of the following is true of hydroforming?
    A. It is used to reinforce the perimeter frame.
    B. It is used for forming large truck frames.
    C. It is done using high pressure.
    D. It is used to make the B pillars.

13. *Technician A* says that the pillars supporting the roof of the car must be able to withstand at least one and a half times the weight of the car. *Technician B* says that the A pillar is sometimes hydroformed. Who is correct?
    A. Technician A only
    B. Technician B only
    C. Both Technicians A and B
    D. Neither Technician A nor B

14. *Technician A* says that lateral stiffeners are sometimes welded into the rocker panel for reinforcement. *Technician B* says that the sail panel is also called the C pillar. Who is correct?
    A. Technician A only
    B. Technician B only
    C. Both Technicians A and B
    D. Neither Technician A nor B

15. *Technician A* says that convolutions are formed into front rail members to reinforce them. *Technician B* says that additional ridges are stamped into the passenger compartment and trunk floor to increase their strength. Who is correct?
    A. Technician A only
    B. Technician B only
    C. Both Technicians A and B
    D. Neither Technician A nor B

Chapter 7

Straightening Steel & Aluminum

■ WORK ASSIGNMENT 7-1

IDENTIFYING AND DESCRIBING AUTOMOTIVE STEEL

Name _____   Date _____

Class _____   Instructor _____   Grade _____

NATEF NONSTRUCTURAL AND DAMAGE REPAIR SECTION II TASKS B. 1, 3

After reading the assignment, in the space provided below, list the personal and environmental safety precautions necessary for this assignment.

1. List the different types of steel used in the manufacturing of a vehicle. Record your findings:

   _____

   _____

   _____

2. What is meant by the terms listed below?
   Yield strength:

   _____

   _____

   _____

   Shear strength:

   _____

   _____

   _____

   Tensile strength:

   _____

   _____

   _____

3. Explain the terms listed below.
   Plasticity of steel:

   _____

   _____

   _____

   Elasticity of steel:

   _____

   _____

   _____

Work hardening of steel:

_____

_____

_____

Direct damage:

_____

_____

_____

Indirect damage:

_____

_____

_____

4. How would you cut, drill, and straighten the steels listed below?

Mild steel:

Cut _____

Drill _____

Straighten _____

High-strength steel:

Cut _____

Drill _____

Straighten _____

Boron steel:

Cut _____

Drill _____

Straighten _____

5. How and why are the types of steel listed above used in automotive manufacturing and why do collision repair technicians need to know about them? Record your findings:

_____

_____

_____

INSTRUCTOR COMMENTS:

_____

_____

_____

_____

_____

_____

■ WORK ASSIGNMENT 7-2

ASSESS DAMAGE AND DEMONSTRATE THE REPAIR TECHNIQUE

Name _____ Date _____

Class _____ Instructor _____ Grade _____

NATEF NONSTRUCTURAL AND DAMAGE REPAIR SECTION II TASKS B. 1, 2, 3, 4, 5, 6; C 1, 2, 3

1. After reading the assignment, in the space provided below, list the personal and environmental safety equipment and precautions needed for this assignment.

_____

_____

_____

2. Explain how a technician would recognize the following:

Direct damage:

_____

_____

Indirect damage:

_____

_____

A crown:

_____

_____

Hinge buckle damage:

_____

_____

Rolled buckle damage:

_____

_____

3. Is it necessary to have access to both sides of a damaged part? If so, why and how should the parts be removed and stored for reattachment?

_____

_____

4. Describe the repair techniques listed below:

Paint removal for metal finishing:

_____

_____

_____

Hammer on dolly:

_____

_____

_____

Hammer off dolly:

_____

_____

_____

Spring hammering:

_____

_____

_____

Pick hammering:

_____

_____

_____

Hammer on spoon:

_____

_____

_____

Weld on dent repair:

_____

_____

_____

Adhesive dent repair:

_____

_____

_____

INSTRUCTOR COMMENTS:

_____

_____

_____

_____

_____

■ WORK ASSIGNMENT 7-3

STRETCHED AND SHRUNKEN METAL AND PAINTLESS DENT REPAIR

Name _____ Date _____

Class _____ Instructor _____ Grade _____

NATEF NONSTRUCTURAL AND DAMAGE REPAIR SECTION II TASKS B. 10; C. 2, 3, 4, 5

1. After reading the assignment, in the space provided below, list the personal and environmental safety equipment and precautions needed for this assignment.

   _____

   _____

   _____

2. Can automotive steel be stretched or shrunken without it fracturing? Explain.

   _____

   _____

   _____

3. How would you identify and determine a repair plan for the following?
   Stretched steel:

   _____

   _____

   Shrunken steel:

   _____

   _____

   _____

   Ripped steel:

   _____

   _____

4. Can shrinking be done cold? If so, explain.

   _____

   _____

5. Can shrinking be done hot? If so, explain.

   _____

   _____

6. What does PDR stand for?

   _____

   _____

   _____

7. Is all damage a candidate for PDR? Explain.

_____

_____

_____

8. How is access important to PDR?

_____

_____

_____

9. Describe the technique of paintless dent repair.

_____

_____

_____

INSTRUCTOR COMMENTS:

_____

_____

_____

_____

_____

_____

ALUMINUM IN AUTOMOTIVE CONSTRUCTION AND COLLISION REPAIR

Name _____   Date _____

Class _____   Instructor _____   Grade _____

NATEF NONSTRUCTURAL AND DAMAGE REPAIR SECTION II TASKS B. 10; C. 2, 3, 4, 5, 6, 7, 8

1. After reading the assignment, in the space provided below, list the personal and environmental safety equipment and precautions needed for this assignment.

   _____

   _____

   _____

2. Describe the steps needed to mix and prepare plastic body filler.

   _____

   _____

   _____

3. Describe the steps needed to apply plastic body filler.

   _____

   _____

   _____

4. Describe the steps needed to sand cured body filler.

   _____

   _____

   _____

5. How has the use of aluminum in automotive construction changed in the last 5 years and how has it affected the collision repair industry?

   _____

   _____

   _____

6. What special precautions must be taken when working on both steel and aluminum?

   In the shop:

   _____

   _____

   _____

   With tools:

   _____

   _____

   _____

   With corrosion:

   _____

   _____

   _____

7. List the types of aluminum used in automotive manufacturing. What are their characteristics?

_____

_____

_____

INSTRUCTOR COMMENTS:

_____

_____

_____

_____

_____

_____

IDENTIFYING AND DESCRIBING AUTOMOTIVE STEEL

Name _____ Date _____

Class _____ Instructor _____ Grade _____

OBJECTIVES

- Know, understand, and use the safety equipment necessary for the task.
- Be able to identify the differing types of steel used in automotive manufacturing.
- Be able to make a repair plan relative to the type of steel that requires repair.
- Be able to explain why heat is used with different types of steel.

NATEF TASK CORRELATION

The written and hands-on activities in this chapter satisfy the NATEF High Priority-Individual and High Priority-Group requirements for NATEF Nonstructural and Damage Repair Section II, Tasks B. 1, 3.

Tools and equipment needed (NATEF tool list)

- Pencil and paper
- Gloves
- Partial mask
- Vehicle repair manual
- Safety glasses
- Ear protection
- MSDS book
- Collision estimating guide (electronic or manual)

Vehicle Description

Year_____ Make _____ Model _____

VIN _____ Paint Code _____

PROCEDURE

1. After reading the work order, gather the safety gear needed to complete the task. In the space provided below, list the personal and environmental safety equipment and precautions needed for this assignment. Have the instructor check and approve your plan before proceeding.

   _____

   _____

   _____

INSTRUCTOR'S APPROVAL _____

Instructions

On a vehicle(s) in the lab, identify the different types of manufacturing material where marked and explain how you reached that conclusion.

2. On the vehicle or parts provided demonstrate how to locate and reduce surface irregularities on a:

   Bent panel:

   _____

   _____

   _____

Kinked panel:

_____

_____

_____

3. How would you straighten high-strength steel that is:
Bent:

_____

_____

_____

Kinked:

_____

_____

_____

4. How would you straighten ultra-high-strength steel that is:
Bent:

_____

_____

_____

Kinked:

_____

_____

_____

5. How would you straighten boron steel that is:
Bent:

_____

_____

_____

Kinked:

_____

_____

_____

6. How does the use of heat differ when repairing the steels listed below?
Mild steel:

_____

_____

_____

High-strength steel:

_____

_____

_____

Boron steel:

_____

_____

_____

INSTRUCTOR COMMENTS:

_____

_____

_____

_____

_____

_____

ASSESS DAMAGE AND DEMONSTRATE THE REPAIR TECHNIQUE

Name _____ Date _____

Class _____ Instructor _____ Grade _____

OBJECTIVES:

- Know, understand, and use the safety equipment necessary for the task.
- Be able to identify different types of damage:
  - Direct damage
  - Indirect damage
  - A crown on a vehicle
  - Hinge buckle
  - Role buckle
- Demonstrate specific repair technique:
  - Hammer on dolly
  - Hammer off dolly
  - Spring hammering
  - Pick hammering
  - Use of a spoon
  - Weld on (stud gun) dent repair tool
  - Adhesive dent repair tool
- Gain working experience in repair techniques.
- Demonstrate gaining access on a vehicle to dent repair.

NATEF TASK CORRELATION

The written and hands-on activities in this chapter satisfy the NATEF High Priority-Individual and High Priority-Group requirements for NATEF Nonstructural and Damage Repair Section II, Task B. 9, 10, 14; C. 2, 3.

Tools and equipment needed (NATEF tool list)

- Pencil and paper
- Gloves
- Partial mask
- Assorted hand tools
- Stud gun with studs

- Safety glasses
- Ear protection
- MSDS book
- Assorted body tools
- Adhesive tent repair tool

Vehicle Description

Year_____ Make _____ Model _____

VIN _____ Paint Code _____

PROCEDURE

1. After reading the work order, gather the safety gear needed to complete the task. In the space provided below, list the personal and environmental safety equipment and precautions needed for this assignment. Have the instructor check and approve your plan before proceeding.

_____

_____

_____

INSTRUCTOR'S APPROVAL _____

2. In the lab, review the damage report and identify the type of damage listed below and write a repair plan for that part.

   Direct damage:

   _____

   _____

   Indirect damage:

   _____

   _____

   Hinge buckle damage:

   _____

   _____

   Rolled buckle damage:

   _____

   _____

3. On a part provided, repair using the techniques below.
   Hammer on dolly:

   Instructor comments:

   _____

   _____

   Hammer off dolly:
   Instructor comments:

   _____

   _____

   Spring hammering:
   Instructor comments:

   _____

   _____

   _____

Pick hammering:
Instructor comments:

_____

_____

_____

Hammer on spoon:
Instructor comments:

_____

_____

_____

Weld-on-dent repair:
Instructor comments:

_____

_____

_____

Adhesive dent repair:
Instructor comments:

_____

_____

_____

INSTRUCTOR COMMENTS:

_____

_____

_____

_____

_____

_____

STRETCHED AND SHRUNKEN METAL

Name _____ Date _____

Class _____ Instructor _____ Grade _____

OBJECTIVES

- Know, understand, and use the safety equipment necessary for the task.
- Identify stretched metal and make a repair plan.
- Identify shrunken metal and make a repair plan.
- Identify ripped metal and make a repair plan.
- Demonstrate repairing stretched metal cold.
- Demonstrate repairing stretched metal hot.

NATEF TASK CORRELATION

The written and hands-on activities in this chapter satisfy the NATEF High Priority-Individual and High Priority-Group requirements for NATEF Nonstructural and Damage Repair Section II, Tasks B. 10; C. 3, 5.

Tools and equipment needed (NATEF tool list)

- Pencil and paper
- Safety glasses
- Gloves
- Ear protection
- Partial mask
- MSDS book
- Assorted body tools
- Heat source (gas or electric)
- Cooling equipment

Vehicle Description

Year_____ Make _____ Model _____

VIN _____ Paint Code _____

PROCEDURE

1. After reading the work order, gather the safety gear needed to complete the task. In the space provided below, list the personal and environmental safety equipment and precautions needed for this assignment. Have the instructor check and approve your plan before proceeding.

   _____

   _____

   _____

INSTRUCTOR'S APPROVAL _____

2. On the vehicle provided, identify and repair stretched metal and write a repair plan.

   _____

   _____

   _____

3. On the vehicle provided, identify and repair shrunken metal and write a repair plan.

_____

_____

_____

4. On the vehicle provided, identify ripped metal and write a repair plan.

_____

_____

_____

5. On a stretched part of the vehicle, repair cold and record your observations.

_____

_____

_____

6. On a stretched part of the vehicle, repair hot and record your observations.

_____

_____

_____

7. Repair a ripped mild steel part.

INSTRUCTOR COMMENTS:

_____

_____

_____

_____

_____

_____

_____

Name _____   Date _____

Class _____   Instructor _____   Grade _____

1. *Technician A* says that thinner body steel is often manufactured by hot rolling. *Technician B* says that the hot-rolling method of manufacturing is commonly done for heavier steel items such as crossmembers. Who is correct?
   A. Technician A only
   B. Technician B only
   C. Both Technicians A and B
   D. Neither Technician A nor B

2. *Technician A* says that the tensile strength of steel is the amount of pulling force that steel can withstand before it deforms or cracks. *Technician B* says that yield strength is the amount of pushing force that steel can withstand before it deforms or cracks. Who is correct?
   A. Technician A only
   B. Technician B only
   C. Both Technicians A and B
   D. Neither Technician A nor B

3. *Technician A* says that plasticity is the process that occurs each time steel is bent. *Technician B* says that plasticity is the ability of steel to take a new shape. Who is correct?
   A. Technician A only
   B. Technician B only
   C. Both Technicians A and B
   D. Neither Technician A nor B

4. *Technician A* says that work hardening occurs when mild steel is heated. *Technician B* says that work hardening is the process used to create Boron steel. Who is correct?
   A. Technician A only
   B. Technician B only
   C. Both Technicians A and B
   D. Neither Technician A nor B

5. *Technician A* says that direct damage is difficult to spot and makes up the majority of the damaged area following a collision. *Technician B* says that direct damage can also be called primary damage. Who is correct?
   A. Technician A only
   B. Technician B only
   C. Both Technicians A and B
   D. Neither Technician A nor B

6. *Technician A* says that indirect damage is the result of the forces of direct damage traveling through the structure. *Technician B* says that it is best to start repairs in the indirect damaged area first. Who is correct?
   A. Technician A only
   B. Technician B only
   C. Both Technicians A and B
   D. Neither Technician A nor B

7. *Technician A* says that hinge buckle often occurs in a crowned area. *Technician B* says that hinge buckle is the result of secondary damage. Who is correct?
   A. Technician A only
   B. Technician B only
   C. Both Technicians A and B
   D. Neither Technician A nor B

8. *Technician A* says that spring hammering is often performed used with a heavy forging hammer. *Technician B* says that hammer off dolly stretches metal. Who is correct?
   A. Technician A only
   B. Technician B only
   C. Both Technicians A and B
   D. Neither Technician A nor B

9. *Technician A* says that stretched metal most often occurs in the indirect damaged area. *Technician B* says that shrinking can only be done by a collision repair technician using the cold method. Who is correct?
   A. Technician A only
   B. Technician B only
   C. Both Technicians A and B
   D. Neither Technician A nor B

10. *Technician A* says that PDR is often performed on large primary damaged areas. *Technician B* says that PDR stands for painted dent repair. Who is correct?
    A. Technician A only
    B. Technician B only
    C. Both Technicians A and B
    D. Neither Technician A nor B

11. *Technician A* says that an aluminum panel can be determined by using a magnet. *Technician B* says that it is not important to distinguish between aluminum and steel panels because the repair techniques and tools are identical. Who is correct?
    A. Technician A only
    B. Technician B only
    C. Both Technicians A and B
    D. Neither Technician A nor B

12. **TRUE** or **FALSE**. One type of aluminum alloy is heat treatable.

Chapter 8

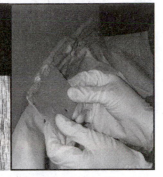

Fillers

■ WORK ASSIGNMENT 8-1

HISTORY, TYPES, AND STORAGE

Name _____   Date _____

Class _____   Instructor _____   Grade _____

NATEF NONSTRUCTURAL AND DAMAGE REPAIR SECTION II TASKS C. 6, 7, 8

1. After reading the assignment, in the space provided below, list the personal and environmental safety precautions necessary for this assignment.

   _____

   _____

   _____

2. Why did plastic body filler replace lead? Record your findings:

   _____

   _____

   _____

3. List the four types of body fillers:

   I _____

   II _____

   III _____

   IV _____

4. What is body filler used for? Record your findings:

   _____

   _____

   _____

5. What is polyester putty used for? Record your findings:

   _____

   _____

   _____

6. What is reinforced filler used for? Record your findings:

   _____

   _____

   _____

7. What is benzyl peroxide used for? Record your findings:

_____

_____

_____

8. How should plastic body filler be stored? Record your findings:

_____

_____

_____

9. How often should containers of stored plastic body filler be "flipped"? Why? Record your findings:

_____

_____

_____

INSTRUCTOR COMMENTS:

_____

_____

_____

_____

_____

_____

PREPARATION, APPLICATION TOOLS, AND MIXING

Name _____   Date _____

Class _____   Instructor _____   Grade _____

NATEF NONSTRUCTURAL AND DAMAGE REPAIR SECTION II TASKS C. 6, 7, 8

1. After reading the assignment, in the space provided below, list the personal and environmental safety precautions necessary for this assignment.

   _____

   _____

   _____

2. Why is surface preparation so important? Record your findings:

   _____

   _____

   _____

3. List the steps for correct surface preparation.

   _____

   _____

   _____

4. Why should the technician cover glass when using a grinder? Record your findings:

   _____

   _____

   _____

5. What grit sandpaper should be used when removing OEM finish? Why?

   _____

   _____

   _____

6. Why is surface preparation so important? Record your findings:

   _____

   _____

   _____

7. How should a plastic applicator be repaired if it gets a nick in it? Record your findings:

   _____

   _____

   _____

8. What PPE should a technician be using when mixing plastic body filler? Record your findings:

   _____

   _____

   _____

9. How much catalyzer should be added to filler? Record your findings:

   _____

   _____

   _____

10. Describe the correct method of mixing body filler. Record your findings:

_____

_____

_____

11. What will overcatalyzing plastic cause? Record your findings:

_____

_____

_____

12. Why is surface preparation so important? Record your findings:

_____

_____

_____

INSTRUCTOR COMMENTS:

_____

_____

_____

_____

_____

_____

SHAPING, SANDING, AND FINISHING MIXING

Name _____   Date _____

Class _____   Instructor _____   Grade _____

NATEF NONSTRUCTURAL AND DAMAGE REPAIR SECTION II TASKS C. 6, 7, 8

1. After reading the assignment, in the space provided below, list the personal and environmental safety precautions necessary for this assignment.

   _____

   _____

   _____

2. Explain the application of the first coat of filler to a flat surface. Record your findings:

   _____

   _____

3. Explain the application of the first coat of filler to a low crowned surface. Record your findings:

   _____

   _____

4. Explain the application of the first coat of filler to a high crown surface. Record your findings:

   _____

   _____

5. Explain the use of a "cheese grater" when shaping plastic body filler. Record your findings:

   _____

   _____

   _____

6. Explain how rough sanding is performed. Record your findings:

   _____

   _____

7. Why is paraffin used in body filler and why must it be removed before an additional coat is applied? Record your findings:

   _____

   _____

   _____

8. What grit paper is used for rough sanding plastic filler? Why? Record your findings:

   _____

   _____

   _____

9. What grit paper should be used for finish sanding and explain how is it done? Record your findings:

_____

_____

_____

10. Explain why and how reinforced filler should be used. Record your findings:

_____

_____

_____

INSTRUCTOR COMMENTS:

_____

_____

_____

_____

_____

_____

FOAMS AND OTHER SPECIALTY FILLERS

Name _____   Date _____

Class _____   Instructor _____   Grade _____

NATEF NONSTRUCTURAL AND DAMAGE REPAIR SECTION II TASKS C. 6, 7, 8

1. After reading the assignment, in the space provided below, list the personal and environmental safety precautions necessary for this assignment.

   _____

   _____

   _____

2. Explain why nonstructural foams are used. Record your findings:

   _____

   _____

   _____

3. Explain how foams control NVH. Record your findings:

   _____

   _____

   _____

4. What is structural foam? How is it used and why? Record your findings:

   _____

   _____

   _____

5. How would a technician know how and which foam to use? Record your findings:

   _____

   _____

   _____

6. Explain what a foam dam is. Record your findings:

   _____

   _____

   _____

7. Explain the application of the first coat of filler to a flat surface. Record your findings:

   _____

   _____

   _____

INSTRUCTOR COMMENTS:

_____

_____

_____

_____

_____

HISTORY, TYPES, AND STORAGE

Name _____   Date _____

Class _____   Instructor _____   Grade _____

OBJECTIVES

- Explain the history of the use of fillers in this industry from its inception to modern day applications.
- Know the many fillers commonly used in the manufacture of an automobile as well as those used when it becomes involved in an accident.
- Be able to select fillers for effecting repairs on specific substrates.
- Understand the composition of fillers and their effect on the longevity of a repair.
- Demonstrate surface preparation and know proper application issues when applying fillers.
- Identify health and safety concerns and PPE requirements relating to the use of fillers and catalysts.
- Identify the tools and equipment typically used for working with plastic and other fillers.
- Answer ASE-style and other review questions and exercises pertaining to fillers.
- Locate and identify specific hazards on MSDS pertaining to filler materials.

NATEF TASK CORRELATION

The written and hands-on activities in this chapter satisfy the NATEF High Priority-Individual and High Priority-Group requirements for Nonstructural and Damage Repair Section II Tasks C. 6, 7, 8.

Tools and equipment needed (NATEF tool list)

- Pencil and paper
- Gloves
- HEPA respirators
- MSDS book
- Body filler
- Polyester putty
- Mixing pallet or board
- Plastic spreaders
- Vehicle description

- Safety glasses
- Ear protection
- Partial mask
- Vehicle repair manual
- Reinforced filler
- Filler catalyst
- Putty knives
- Paper towels
- Collision estimating guide (electronic or manual)

Vehicle Description

Year_____ Make _____ Model _____

VIN _____ Paint Code _____

PROCEDURE

1. After reading the work order, gather the safety gear needed to complete the task. In the space provided below, list the personal and environmental safety equipment and precautions needed for this assignment. Have the instructor check and approve your plan before proceeding.

_____

_____

_____

INSTRUCTOR'S APPROVAL _____

On a vehicle or part provided:

1. Choose the body filler application tool you will use and have it nearby.

2. Determine the correct amount of plastic body filler and place it on the pallet. Record your steps:

   _____

   _____

   _____

3. Determine the correct amount of catalyst for the body filler and place it on the filler. Record your steps:

   _____

   _____

   _____

4. Mix the filler until no streaks are visible. Record your steps:

   _____

   _____

   _____

5. Apply the filler. Record your steps:

   _____

   _____

   _____

INSTRUCTOR COMMENTS:

_____

_____

_____

_____

_____

_____

PREPARATION, APPLICATION TOOLS, AND MIXING

Name _____   Date _____

Class _____   Instructor _____   Grade _____

OBJECTIVES

- Explain the history of the use of fillers in this industry from its inception to modern day applications.
- Know the many fillers commonly used in the manufacture of an automobile as well as those used when it becomes involved in an accident.
- Be able to select fillers for effecting repairs on specific substrates.
- Understand the composition of fillers and their effect on the longevity of a repair.
- Demonstrate surface preparation and know proper application issues when applying fillers.
- Identify health and safety concerns and PPE requirements relating to the use of fillers and catalysts.
- Identify the tools and equipment typically used for working with plastic and other fillers.
- Answer ASE-style and other review questions and exercises pertaining to fillers.
- Locate and identify specific hazards on MSDS pertaining to filler materials.

NATEF TASK CORRELATION

The written and hands-on activities in this chapter satisfy the NATEF High Priority-Individual and High Priority-Group requirements for Nonstructural and Damage Repair Section II Tasks C. 6, 7, 8.

Tools and equipment needed (NATEF tool list)

- Pencil and paper
- Gloves
- HEPA respirators
- MSDS book
- Body filler
- Reinforced filler
- Filler catalyst
- Putty knives
- Paper towels

- Safety glasses
- Ear protection
- Partial mask
- Vehicle repair manual
- Collision estimating guide (electronic or manual)
- Polyester putty
- Mixing pallet or board
- Plastic spreaders
- Vehicle description

Vehicle Description

Year_____ Make _____ Model _____

VIN _____ Paint Code _____

PROCEDURE

1. After reading the work order, gather the safety gear needed to complete the task. In the space provided below, list the personal and environmental safety equipment and precautions needed for this assignment. Have the instructor check and approve your plan before proceeding.

_____

_____

_____

INSTRUCTOR'S APPROVAL _____

On a vehicle or part provided:

2. Using a surform grater shape the partially cured plastic. Record your steps:

_____

_____

_____

3 Using the correct grit sandpaper rough sand the filler. Record your steps:

_____

_____

_____

4. Using the correct grit sandpaper finish sand the filler. Record your steps:

_____

_____

_____

5. Apply additional coats of filler as needed. Record your steps:

_____

_____

_____

INSTRUCTOR COMMENTS:

_____

_____

_____

_____

_____

SHAPING, SANDING, AND FINISHING MIXING

Name _____   Date _____

Class _____   Instructor _____   Grade _____

OBJECTIVES

- Explain the history of the use of fillers in this industry from its inception to modern day applications.
- Know the many fillers commonly used in the manufacture of an automobile as well as those used when it becomes involved in an accident.
- Be able to select fillers for effecting repairs on specific substrates.
- Understand the composition of fillers and their effect on the longevity of a repair.
- Demonstrate surface preparation and know proper application issues when applying fillers.
- Identify health and safety concerns and PPE requirements relating to the use of fillers and catalysts.
- Identify the tools and equipment typically used for working with plastic and other fillers.
- Answer ASE-style and other review questions and exercises pertaining to fillers.
- Locate and identify specific hazards on MSDS pertaining to filler materials.

NATEF TASK CORRELATION

The written and hands-on activities in this chapter satisfy the NATEF High Priority-Individual and High Priority-Group requirements for Nonstructural and Damage Repair Section II Tasks C. 6, 7, 8.

Tools and equipment needed (NATEF tool list)

- Pencil and paper
- Gloves
- HEPA respirators
- MSDS book
- Body filler
- Reinforced filler
- Filler catalyst
- Putty knives
- Paper towels

- Safety glasses
- Ear protection
- Partial mask
- Vehicle repair manual
- Collision estimating guide (electronic or manual)
- Polyester putty
- Mixing pallet or board
- Plastic spreaders

Vehicle Description

Year_____ Make _____ Model _____

VIN _____ Paint Code _____

PROCEDURE

1. After reading the work order, gather the safety gear needed to complete the task. In the space provided below, list the personal and environmental safety equipment and precautions needed for this assignment. Have the instructor check and approve your plan before proceeding.

_____

_____

_____

INSTRUCTOR'S APPROVAL _____

2. In an area designated by your instructor, while using the correct PPE, inject about ½ oz of nonstructural foam into a cup. Record your observations:

_____

_____

_____

3. After the foam has cured, cut it in half and observe the cell structure. Explain how this type of foam could reduce NVH:

_____

_____

_____

4. In an area designated by your instructor, while using the correct PPE, inject about ½ oz of structural foam into a cup. Record your observations:

_____

_____

_____

5. After the foam has cured, cut it in half and observe the cell structure. Explain how this type of foam could increase structural integrity.

_____

_____

_____

INSTRUCTOR COMMENTS:

_____

_____

_____

_____

_____

_____

Name _____   Date _____

Class _____   Instructor _____   Grade _____

1. *Technician A* says that all plastic fillers and fiberglass fillers are catalyzed using the same butyl benzoyl peroxide hardener. *Technician B* says that most plastic fillers use an epoxy base resin. Who is correct?
   A. Technician A only
   B. Technician B only
   C. Both Technicians A and B
   D. Neither Technician A nor B

2. *Technician A* says that clean cardboard can be used for a plastic mixing surface. *Technician B* says that plastic fillers can readily be applied over a rust-perforated surface without consequence as long as the scale has been removed. Who is correct?
   A. Technician A only
   B. Technician B only
   C. Both Technicians A and B
   D. Neither Technician A nor B

3. *Technician A* says that a can of plastic filler can be put into a paint shaker to thoroughly agitate it. *Technician B* says that the liquid-like material that collects on top of the can of filler is responsible for its adhesion to the surface. Who is correct?
   A. Technician A only
   B. Technician B only
   C. Both Technicians A and B
   D. Neither Technician A nor B

4. *Technician A* says that structural foam is used to reduce the NVH on the vehicle. *Technician B* says that repair mapping may occur if plastic is overcatalyzed. Who is correct?
   A. Technician A only
   B. Technician B only
   C. Both Technicians A and B
   D. Neither Technician A nor B

5. *Technician A* says that grating the plastic filler will reveal the low spots on the surface when the excess material that was applied has been removed. *Technician B* follows up the grating step with 80-grade sandpaper. Who is correct?
   A. Technician A only
   B. Technician B only
   C. Both Technicians A and B
   D. Neither Technician A nor B

6. *Technician A* says that the aluminum clad body fillers require the use of liquid hardener. *Technician B* says that the premium grade fillers are less apt to stain than the conventional fillers. Who is correct?
   A. Technician A only
   B. Technician B only
   C. Both Technicians A and B
   D. Neither Technician A nor B

7. *Technician A* says that a two-part glaze coat should be used as a final finish coat over the entire repair area. *Technician B* says that the paint should be removed approximately 3 to 4 inches beyond the edge of the repaired surface when using conventional or lightweight fillers. Who is correct?
   A. Technician A only
   B. Technician B only
   C. Both Technicians A and B
   D. Neither Technician A nor B

8. *Technician A* says that automobile manufacturers use stuffers inside doors and other cavities to help reduce the NVH. *Technician B* says that most manufacturers use preshaped foam to help absorb collision energy. Who is correct?
   A. Technician A only
   B. Technician B only
   C. Both Technicians A and B
   D. Neither Technician A nor B

9. *Technician A* says that structural foams are used in the areas of the vehicle's "A pillars" and "D pillars." *Technician B* says that the only place structural foam is used is on the lower rails and in the torque box area of some frame rails. Who is correct?
   A. Technician A only
   B. Technician B only
   C. Both Technicians A and B
   D. Neither Technician A nor B

10. *Technician A* says that visible streaks of catalyst or hardener left in the plastic when applying it can cause staining problems. *Technician B* says most plastic filler manufacturers recommend a ratio of approximately 2% hardener to the total plastic batch. Who is correct?
    A. Technician A only
    B. Technician B only
    C. Both Technicians A and B
    D. Neither Technician A nor B

Chapter 9

Trim & Hardware

■ WORK ASSIGNMENT 9-1

LABEL AND STORE REMOVED PARTS, TRIM, AND HARDWARE

Name _____ Date _____

Class _____ Instructor _____ Grade _____

NATEF NONSTRUCTURAL ANALYSIS AND DAMAGE REPAIR SECTION II A. 1, 2, 3, 4, 5, 6, 7, 9

1. After reading the assignment, in the space provided below, list the personal and environmental safety equipment and precautions needed for this assignment.

   _____

   _____

   _____

2. List why it is important for technicians to label fasteners when they are removed.

   _____

   _____

   _____

3. Why is proper storage helpful for reducing reassembly time?

   _____

   _____

   _____

4. What tools and equipment will be needed to label and store fasteners and hardware when removed?

   _____

   _____

   _____

5. What is torque and how does it relate to fasteners?

   _____

   _____

   _____

6. What does SAE stand for?

   _____

   _____

7. If a bolt that has a 4.8 marking in the head is removed, is it an SAE bolt or a metric bolt? What indicates it?

   _____

   _____

8. How would you identify a grade 5 ASE bolt?

_____

_____

9. Explain what is meant by single-use fasteners and give examples.

_____

_____

10. What is the purpose of a flat washer?

_____

_____

11. What is the purpose of a split lockwasher?

_____

_____

12. What is meant by an SPR fastener?

_____

_____

13. As trim is removed, if the fasteners break, what should be done before reassembly?

_____

_____

14. Why is it important to replace the identical appearing fasteners when reassembling a repaired vehicle?

_____

_____

15. What is the difference between a flat washer and a flat fender washer?

_____

_____

16. What is meant by a single-use fastener and where may it be used?

_____

_____

17. How can a technician distinguish between a fastener that is intended for single use and other fasteners?

_____

_____

_____

INSTRUCTOR COMMENTS:

_____

_____

_____

_____

■ WORK ASSIGNMENT 9-2

REMOVE TRIM AND HARDWARE

Name _____ Date _____

Class _____ Instructor _____ Grade _____

NATEF NONSTRUCTURAL ANALYSIS AND DAMAGE REPAIR SECTION II A. 1, 2, 3, 4, 5, 6, 7, 9

1. After reading the assignment, in the space provided below, list the personal and environmental safety equipment and precautions needed for this assignment.

   _____

   _____

   _____

2. On a vehicle provided in the lab, read the repair order and make a plan to remove the vehicle's identification bagging. Record the removal plan you have made.

   _____

   _____

   _____

3. Which trim removal tools will you need for the job?

   _____

   _____

   _____

4. Explain how you would clean and inspect to evaluate if it can be reinstalled; if serviceable, label and store. Record your observations.

   _____

   _____

   _____

5. Prepare the bag for reinstallation. Record the process.

   _____

   _____

   _____

6. Describe how you would reinstall it in its proper location. Record the process.

   _____

   _____

   _____

7. On a vehicle provided in the lab, explain how a technician would remove the vehicle's belt molding. Record the removal plan you have made.

   _____

   _____

   _____

8. Which trim removal tools will you need for the job?

   _____

   _____

   _____

9. Describe the process to clean and inspect it to evaluate if it can be reinstalled; if serviceable label and store. Record your observations.

_____

_____

_____

10. List the steps to prepare belt molding for reinstallation. Record the process.

_____

_____

_____

11. Describe how to reinstall the belt in the proper location. Record the process.

_____

_____

_____

12. On a vehicle provided in the lab, explain how to remove the vehicle's quarter cladding molding. Record the removal plan you have made.

_____

_____

_____

13. Which trim removal tools will you need for the job?

_____

_____

_____

14. List the steps to clean then inspect to evaluate if it can be reinstalled; if serviceable, label and store. Record your observations.

_____

_____

_____

15. How would a technician prepare quarter cladding molding for reinstallation? Record the process.

_____

_____

_____

16. List the steps to reinstall the cladding in its proper location. Record the process.

_____

_____

_____

INSTRUCTOR COMMENTS:

_____

_____

_____

_____

_____

_____

■ WORK ASSIGNMENT 9-3

REMOVE AND INSTALL MOLDING WEATHERSTRIPPING

Name _____ Date _____

Class _____ Instructor _____ Grade _____

NATEF NONSTRUCTURAL ANALYSIS AND DAMAGE REPAIR SECTION II A. 1, 2, 3, 4, 5, 6, 7, 9

1. After reading the assignment, in the space provided below, list the personal and environmental safety equipment and precautions needed for this assignment.

   _____

   _____

   _____

2. On a vehicle provided, read the damage report and make a repair plan to remove the front seat, including the seat belt restraint. List the process.

   _____

   _____

   _____

3. Which tools will you need for the job?

   _____

   _____

4. Explain how you would clean and inspect to evaluate if it can be reinstalled; if serviceable, label and store. Record your observations.

   _____

   _____

   _____

5. List the steps to prepare it for reinstallation. Record the process.

   _____

   _____

6. Describe how a technician would reinstall it in its proper location. Record the process.

   _____

   _____

   _____

7. On a vehicle provided, make a repair plan to remove the driver's door (both inner and outer) weatherstripping. List the process.

   _____

   _____

   _____

8. Which tools will you need for the job?

   _____

   _____

   _____

9. Explain how you would clean and inspect to evaluate if it can be reinstalled. Record your observations.

_____

_____

_____

10. List the steps to prepare it for reinstallation. Record the process.

_____

_____

_____

11. Describe how a technician would reinstall it in its proper location. Record the process.

_____

_____

_____

12. On a vehicle provided, make a repair plan to remove the driver's side body molding. List the process.

_____

_____

_____

13. Which tools will you need for the job?

_____

_____

_____

14. Explain how you would clean and inspect to evaluate if it can be reinstalled. Record your observations.

_____

_____

_____

15. List the steps to prepare it for reinstallation. Record the process.

_____

_____

_____

16. Describe how a technician would reinstall it in its proper location. Record the process.

_____

_____

_____

INSTRUCTOR COMMENTS:

_____

_____

_____

_____

_____

_____

LABEL AND STORE REMOVED PARTS, TRIM, AND HARDWARE

Name _____   Date _____

Class _____   Instructor _____   Grade _____

OBJECTIVES

- Know, understand, and use the safety equipment necessary for the task.
- Label and store removed trim, moldings, hardware, and other parts for reinstallation following the repair of a vehicle.
- Be able to identify and properly remove or reinstall the more common bolts, fasteners, screws, and retainers.

NATEF TASK CORRELATION

The written and hands-on activities in this chapter satisfy the NATEF High Priority-Individual and High Priority-Group requirements for nonstructural analysis and damage repair Section II A. 1, 2, 3, 4, 5, 6, 7, 9.

Tools and equipment needed (NATEF tool list)

- Safety glasses
- Gloves
- Ear protection
- Particle mask
- Pencil and paper
- Plastic bags
- Storage cart
- Assorted hand tools
- Trim-removing tools

Vehicle Description

Year_____ Make _____ Model _____

VIN _____ Paint Code _____

PROCEDURE

1. After reading the work order, gather the safety gear needed to complete the task. In the space provided below, list the personal and environmental safety equipment and precautions needed for this assignment. Have the instructor check and approve your plan before proceeding.

_____

_____

_____

INSTRUCTOR'S APPROVAL _____

2. On a vehicle provided, remove the front bumper and label and store the fasteners. Record your activity:

_____

_____

_____

3. List any specialty tool needed for removal.

_____

_____

_____

4. Inspect the removed fasteners and identify any that may not be suitable for reuse. Record your findings:

_____

_____

_____

5. On a vehicle provided, reinstall the front bumper using the stored fastener. Record your activity:

_____

_____

_____

6. On a vehicle provided, remove the interior door trim and pad to service the window regulator. Label and store the fasteners and parts. Record your activity:

_____

_____

_____

7. List any specialty tool needed for removal.

_____

_____

_____

8. Inspect the removed fasteners and identify any that may not be suitable for reuse. Record your findings:

_____

_____

_____

9. On a vehicle provided, reinstall the interior door trim and pad. Record your activity:

_____

_____

_____

10. On a vehicle provided, remove the fender and label and store the fasteners and parts for reinstallation. Record your activity:

_____

_____

_____

11. List any specialty tool needed for removal.

_____

_____

_____

12. Inspect the removed fasteners and identify any that may not be suitable for reuse. Record your findings:

_____

_____

_____

13. On a vehicle provided, reinstall the fender and align it. Record your activity:

_____

_____

_____

14. Have the vehicle inspected by the instructor for final evaluation.

_____

_____

_____

INSTRUCTOR COMMENTS:

_____

_____

_____

_____

_____

_____

REMOVE AND INSTALL MOLDING WEATHERSTRIPPING

Name _____ Date _____

Class _____ Instructor _____ Grade _____

OBJECTIVES

- Know, understand, and use the safety equipment necessary for the task.
- Remove and reinstall common interior trim, molding, and weatherstripping.
- Remove and reinstall interior components.

NATEF TASK CORRELATION

The written and hands-on activities in this chapter satisfy the NATEF High Priority-Individual and High Priority-Group requirements for nonstructural analysis and damage repair Section II A. 1, 2, 3, 4, 5, 6, 7, 9.

Tools and equipment needed (NATEF tool list)

- Safety glasses
- Ear protection
- Pencil and paper
- Storage cart
- Trim-removing tools
- Mounting tape
- Gloves
- Particle mask
- Plastic bags
- Assorted hand tools
- Adhesive
- ¾ inch masking tape

Vehicle Description

Year_____ Make _____ Model _____

VIN _____ Paint Code _____

PROCEDURE

1. After reading the work order, gather the safety gear needed to complete the task. In the space provided below, list the personal and environmental safety equipment and precautions needed for this assignment. Have the instructor check and approve your plan before proceeding.

   _____

   _____

   _____

INSTRUCTOR'S APPROVAL _____

2. On a vehicle provided, remove the front seat, including the seat belt restraint; label and store the fasteners. Record your activity:

   _____

   _____

   _____

3. List any specialty tool needed for removal.

   _____

   _____

   _____

4. Inspect the removed fasteners to identify if they are suitable for reuse. Record your findings:

_____

_____

_____

5. On a vehicle provided, reinstall the front seat, including the seat belt restraint using the stored fasteners or new adhesive. Record your activity:

_____

_____

_____

6. On a vehicle provided, remove the driver's door (both inner and outer) weatherstripping and store them and their fasteners. Record your activity:

_____

_____

_____

7. List any specialty tool needed for removal.

_____

_____

_____

8. Inspect the removed fasteners to identify if they are suitable for reuse. Record your findings:

_____

_____

_____

9. On a vehicle provided, reinstall the driver's door (both inner and outer) weatherstripping, using the stored fasteners or new adhesive. Record your activity:

_____

_____

_____

10. On a vehicle provided, remove the driver's side body molding and store them and their fasteners. Record your activity:

_____

_____

_____

11. List any specialty tool needed for removal.

_____

_____

_____

12. Inspect the removed fasteners to identify if they are suitable for reuse. Record your findings:

_____

_____

_____

13. On a vehicle provided, reinstall the driver's side body molding using the stored fasteners or new adhesive. Record your activity:

_____

_____

_____

INSTRUCTOR COMMENTS:

_____

_____

_____

_____

_____

_____

Name _____ Date _____

Class _____ Instructor _____ Grade _____

1. *Technician A* says that it is vital for correct reinstallation to properly label and store parts and fasteners when removed. *Technician B* says that with the convenience of cell phone cameras, a technician can photograph parts, their fasteners, and electrical connections for reconnection. Who is correct?
   A. Technician A only
   B. Technician B only
   C. Both Technicians A and B
   D. Neither Technician A nor B

2. *Technician A* says that specially designed equipment is needed to efficiently label and store trim and hardware. *Technician B* says that masking tape, zip lock bags, and a pen can efficiently be used to label and store most fasteners. Who is correct?
   A. Technician A only
   B. Technician B only
   C. Both Technicians A and B
   D. Neither Technician A nor B

3. *Technician A* says that all threaded fasteners have a torque although some may be more critical that a precise torque is applied. *Technician B* says that a torque value only applies to large major fasteners. Who is correct?
   A. Technician A only
   B. Technician B only
   C. Both Technicians A and B
   D. Neither Technician A nor B

4. *Technician A* says that a bolt with three slash marks on its head indicates that it is an SAE bolt grade 5. *Technician B* says that a bolt with three slash marks on its head indicates that it is a metric bolt grade 3. Who is correct?
   A. Technician A only
   B. Technician B only
   C. Both Technicians A and B
   D. Neither Technician A nor B

5. *Technician A* says that single-use fasteners are only found on mechanical parts such as suspension. *Technician B* says that single-use fastener identification is best found in the vehicle service manual. Who is correct?
   A. Technician A only
   B. Technician B only
   C. Both Technicians A and B
   D. Neither Technician A nor B

6. *Technician A* says that one of the purposes of vehicle cladding is to prevent stone chipping on the lower part of the vehicle. *Technician B* says that cladding is purely cosmetic and does not prevent chipping. Who is correct?
   A. Technician A only
   B. Technician B only
   C. Both Technicians A and B
   D. Neither Technician A nor B

7. *Technician A* says that many vehicle moldings and nameplate bags can be removed with fishing line. *Technician B* says that many vehicle moldings and nameplate bags can be removed with low heat. Who is correct?
   A. Technician A only
   B. Technician B only
   C. Both Technicians A and B
   D. Neither Technician A nor B

8. **TRUE** or **FALSE**. To properly sand and prepare the vehicle's surface for paint, trim must be removed and replaced (R&R).

9. *Technician A* says that seat belt fasteners may be secured with thread glue, thus making them difficult to remove. *Technician B* says that thread glue can be softened with heat. Who is correct?
   A. Technician A only
   B. Technician B only
   C. Both Technicians A and B
   D. Neither Technician A nor B

10. *Technician A* says that fasteners whose threads have been damaged when removed may need to be replaced. *Technician B* says that if a fastener was removed and found to have thread adhesive on it, it must be replaced with a new fastener. Who is correct?
    A. Technician A only
    B. Technician B only
    C. Both Technicians A and B
    D. Neither Technician A nor B

ASE-STYLE REVIEW QUESTIONS

1. *Technician A* says that self-tapping screws are used to hold sheet metal together. *Technician B* says that self-tapping screws come as Phillips heads only. Who is correct?
   A. Technician A only
   B. Technician B only
   C. Both Technicians A and B
   D. Neither Technician A nor B

2. *Technician A* says that cladding is the name of the interior trim of automobiles. *Technician B* says that weatherstripping is always attached with adhesives. Who is correct?
   A. Technician A only
   B. Technician B only
   C. Both Technicians A and B
   D. Neither Technician A nor B

3. *Technician A* says that bagging and tagging is the covering of a vehicle to protect it from overspray. *Technician B* says that the proper labeling and storing of removed parts is vitally important to the reassembly after repairs. Who is correct?
   A. Technician A only
   B. Technician B only
   C. Both Technicians A and B
   D. Neither Technician A nor B

4. *Technician A* says that the use of protective mechanic's gloves is important when removing trim and is part of the personal protective clothing of a repair technician. *Technician B* says that safety glasses should be worn at all times when working in a collision repair shop. Who is correct?
   A. Technician A only
   B. Technician B only
   C. Both Technicians A and B
   D. Neither Technician A nor B

5. *Technician A* says that a bolt is usually used to hold together two removable parts. *Technician B* says that SAE stands for "Society of American Engineers." Who is correct?
   A. Technician A only
   B. Technician B only
   C. Both Technicians A and B
   D. Neither Technician A nor B

6. *Technician A* says that "a ¼ -20 × 3" is a ¼-inch diameter fastener with 20 threads per inch that is 3 inches long. *Technician B* says that to determine the pitch of a fastener a thread-pitch gauge should be used. Who is correct?
   A. Technician A only
   B. Technician B only
   C. Both Technicians A and B
   D. Neither Technician A nor B

7. *Technician A* says that bolts come in both right- and left-handed thread. *Technician B* says that to loosen a right-handed thread bolt, it should be turned to the right. Who is correct?
   A. Technician A only
   B. Technician B only
   C. Both Technicians A and B
   D. Neither Technician A nor B

8. *Technician A* says that flat washers are used to spread out the clamping power of the fastener. *Technician B* says that pop rivets are also referred to as blind rivets. Who is correct?
   A. Technician A only
   B. Technician B only
   C. Both Technicians A and B
   D. Neither Technician A nor B

9. *Technician A* says that an SPR is a self-punching rivet. *Technician B* says that to "R&I" a part means that the technician will repair and install a part. Who is correct?
   A. Technician A only
   B. Technician B only
   C. Both Technicians A and B
   D. Neither Technician A nor B

10. *Technician A* says that a castle nut uses a cotter pin to ensure that it stays tight. *Technician B* says that snaprings are fasteners that fit into a groove on the inside of a part. Who is correct?
    A. Technician A only
    B. Technician B only
    C. Both Technicians A and B
    D. Neither Technician A nor B

11. *Technician A* says that if a bolt has a grade of 4.6 marked on its head, it is a metric bolt. *Technician B* says that grade 8 bolts can be replaced with grade 5 bolts if needed. Who is correct?
    A. Technician A only
    B. Technician B only
    C. Both Technicians A and B
    D. Neither Technician A nor B

12. *Technician A* says that nameplates can be removed with fishing line. *Technician B* says that nameplates can be removed with heat and a plastic putty knife. Who is correct?
    A. Technician A only
    B. Technician B only
    C. Both Technicians A and B
    D. Neither Technician A nor B

13. *Technician A* says that appliqués can be either painted on or glued on. *Technician B* says that an eraser disk should be turned fast to be effective. Who is correct?
    A. Technician A only
    B. Technician B only
    C. Both Technicians A and B
    D. Neither Technician A nor B

14. All except which one of these could be considered to be exterior trim?
    A. cladding
    B. belt molding
    C. weatherstripping
    D. moldings

15. *Technician A* says that decals should be removed with a razor blade. *Technician B* says that decals can be installed either wet or dry. Who is correct?
    A. Technician A only
    B. Technician B only
    C. Both Technicians A and B
    D. Neither Technician A nor B

ESSAY-STYLE REVIEW QUESTIONS

1. Explain why parts should be labeled and stored when removed from a vehicle during repair.

   _____

   _____

   _____

2. Analyze and describe how SAE bolt grading can be recognized by a technician.

   _____

   _____

   _____

3. Explain why it is important to inform the parts department that clips or fasteners are damaged when removing parts from a vehicle.

_____

_____

4. Describe the procedure used to remove adhesive-bonded nameplates with "fishing line."

_____

_____

5. Describe how to make a template from tape before removing a nameplate.

_____

_____

6. Explain how a self-locking nut works.

_____

_____

7. Explain why weatherstripping is needed around a door.

_____

_____

LABORATORY ACTIVITIES

1. In the lab, set up an assortment of fasteners that the students must identify and label.
2. Have students remove a vehicle's door and disassemble it, then reassemble it and reinstall it on the vehicle.
3. Have students remove a body side molding, clean the vehicle and molding of adhesive residue, then reinstall the molding on the vehicle.
4. Have students prepare a vehicle for refinishing by de-trimming the vehicle: removing side mirror, belt molding, body side molding, door handle and lock, appliqué, and any other trim that would interfere with refinishing.
5. Have students remove the interior of a vehicle (seats, carpet, and needed trim), then reinstall.

TOPIC-RELATED MATH QUESTIONS

1. A technician is removing a weatherstripping that has 33 plastic fasteners. 50% of the fasteners break during removal. How many fasteners must the technician purchase? (16.5)
2. The parts department uses 32 fender fasteners a week. How many fasteners would the department use in a year? (1664)
3. If a bolt is 2½ inches long, how many millimeters long would it be? (63.5 mm)
4. If two-sided adhesive comes in a 7-yard roll that costs $26.34, how much does it cost per foot? ($1.25/foot)
5. If you bought two-sided adhesive for $3.00 per foot and you marked it up 150%, how much would you charge for a yard of it? ($22.50)

CRITICAL THINKING QUESTIONS

1. After reading this chapter, explain the importance of the parts department in a body shop.

   _____

   _____

   _____

2. Why is a cotter key a single-use fastener?

   _____

   _____

   _____

3. Why should a collision repair be an undetectable repair?

   _____

   _____

   _____

Chapter 10

Estimating Collision Damage

■ WORK ASSIGNMENT 10-1

DAMAGE REPORTS

Name _____ Date _____

Class _____ Instructor _____ Grade _____

1. After reading the assignment, in the space provided below, list the personal and environmental safety equipment and precautions needed for this assignment.

   _____

   _____

   _____

2. List other names for a damage report that are commonly used in the industry. Record your findings:

   _____

   _____

   _____

3. Describe how a manual estimate is written. Record your findings:

   _____

   _____

   _____

4. Describe how a computerized estimate is written and its advantages. Recorded your findings:

   _____

   _____

   _____

5. How does digital imaging help create a damage report? Record your findings:

   _____

   _____

   _____

6. List the information that should be gathered from the owner before writing the estimate.
   Items:

   _____

   _____

   _____

7. List what an estimator should do on initial vehicle inspection. Record your findings:

   _____

   _____

   _____

8. Why should a set sequence of inspection be followed? Record your findings:

_____

_____

_____

9. Explain what a supplement is and why it may be necessary to write one even on a well-prepared initial damage report. Record your findings:

_____

_____

_____

INSTRUCTOR COMMENTS:

_____

_____

_____

_____

_____

_____

DAMAGE REPORT TERMS

Name _____ Date _____

Class _____ Instructor _____ Grade _____

1. After reading the assignment, in the space provided below, list the personal and environmental safety equipment and precautions needed for this assignment.

   _____

   _____

   _____

Types of insurance coverage

2. Describe the different types of insurance coverage listed below.

   Liability:

   _____

   _____

   Collision:

   _____

   _____

   Comprehensive:

   _____

   _____

   Deductibles:

   _____

   _____

   Direct repair:

   _____

   _____

Estimating terms

3. Give a short definition or explanation of the terms listed below.

   Betterment:

   _____

   _____

   Depreciation:

   _____

   _____

Appearance allowance:

_____

_____

_____

R & R:

_____

_____

R & I:

_____

_____

O/H:

_____

_____

IOH:

_____

_____

O/L:

_____

_____

AT:

_____

_____

PED:

_____

_____

SUB:

_____

_____

VIN:

_____

_____

_____

INSTRUCTOR COMMENTS:

_____

_____

_____

_____

_____

DAMAGE REPORTS

Name _____   Date _____

Class _____   Instructor _____   Grade _____

OBJECTIVES

- Define the estimate and damage report types, including:
  - Manual
  - Computer
  - Digital imaging
- See how estimates are made and what they are utilized for, including:
  - Sales
  - Parts orders
  - Repair orders
- Understand various facets of insurance coverage and programs, including:
  - Liability
  - Collision
  - Comprehensive
  - Deductibles
  - Direct repair
- Utilize and understand industry terms, including:
  - Betterment
  - Depreciation
  - Appearance allowance
  - R&I (remove and install)
  - R&R (remove and replace)
  - O/H (overhaul)
  - IOH (included in overhaul)
  - O/L (overlap)
  - AT (access time)
  - PED (preexisting damage)
  - SUB (sublet)
- Identify a vehicle, including:
  - Decoding a VIN
  - Refinish, trim, and various other codes
- Identify and analyze types of damage to a vehicle, including:
  - Visual damage indicators
  - Measuring for damage
  - Primary and secondary damage
  - Repair or replace?
  - Plastics and other substrates
  - The different areas of a vehicle (subject to damages)
- Utilize estimating guides and understand their facets, including:
  - Parts information
  - Labor information
  - Procedural explanations
  - Refinish operations
  - Price information
- Evaluate a vehicle's worth, and determine whether or not it is a:
  - Total loss
  - Repairable vehicle

NATEF TASK CORRELATION

The written and hands-on activities in this chapter satisfy the NATEF High Priority-Individual and High Priority-Group requirements. Though there are no NATEF requirements for this section, it is felt that a working knowledge of the industry is necessary.

Tools and equipment needed (NATEF tool list)
- Pen and pencil
- Safety glasses
- Gloves
- Estimating manuals
- Computerized estimating equipment
- Digital camera
- Assorted hand tools

Instructions

The activities for this section are intended to acquaint you with the hand tools. Your instructor will train each of you on the safe use of the tools that you will identify. Safety is a primary concern and you must follow all instructions, both written and demonstrated, and should not operate equipment or use tools without the expressed direction of your instructor.

PROCEDURE

In the lab, after completing the work assignment for that section, follow the work order and record your findings for your instructor to review. If the section calls for "instructor approval," you should not proceed without this approval.

1. After reading the work order, gather the safety gear needed to complete the task. In the space provided below, list the personal and environmental safety equipment and precautions needed for this assignment. Have the instructor check and approve your plan before proceeding.

_____

_____

_____

INSTRUCTOR'S APPROVAL _____

2. In the lab using the vehicle provided, gather the tools and information needed for:
   Vehicle description

   Owner's name _____

   Address _____

   Home phone _____

   Cell number _____

   e-mail address _____

   Insurance company_____

   Adjuster's name _____

   Phone: Work _____ Cell_____

   Year _____ Make _____ Model _____

   VIN _____ Production Date _____ OEM Paint Code_____

   Area of direct damage _____

   Photo Number/s _____

3. List the tools and equipment that will be necessary to estimate this vehicle manually. Record your findings:

_____

_____

_____

4. List the tools and equipment that will be necessary to estimate this vehicle using a computer. Record your findings:

_____

_____

_____

5. How would digital photographs help when estimating a damaged vehicle? Record your findings:

_____

_____

_____

6. Where was this vehicle manufactured?

_____

7. Where did you find the production date?

_____

8. Where did you find the OEM paint code?

_____

INSTRUCTOR COMMENTS:

_____

_____

_____

_____

_____

_____

DAMAGE REPORTS (MANUAL)

Name _____   Date _____

Class _____   Instructor _____   Grade _____

OBJECTIVES

- Define the estimate and damage report types, including:
  - Manual
  - Computer
  - Digital imaging
- See how estimates are made and what they are utilized for, including:
  - Sales
  - Parts orders
  - Repair orders
- Understand various facets of insurance coverage and programs, including:
  - Liability
  - Collision
  - Comprehensive
  - Deductibles
  - Direct repair
- Utilize and understand industry terms, including:
  - Betterment
  - Depreciation
  - Appearance allowance
  - R&I (remove and install)
  - R&R (remove and replace)
  - O/H (overhaul)
  - IOH (included in overhaul)
  - O/L (overlap)
  - AT (access time)
  - PED (preexisting damage)
  - SUB (sublet)
- Identify a vehicle, including:
  - Decoding a VIN
  - Refinish, trim, and various other codes
- Identify and analyze types of damage to a vehicle, including:
  - Visual damage indicators
  - Measuring for damage
  - Primary and secondary damage
  - Repair or replace?
  - Plastics and other substrates
  - The different areas of a vehicle (subject to damages)
- Utilize estimating guides and understand their facets, including:
  - Parts information
  - Labor information
  - Procedural explanations
  - Refinish operations
  - Price information
- Evaluate a vehicle's worth, and determine whether or not it is a:
  - Total loss
  - Repairable vehicle

NATEF TASK CORRELATION

The written and hands-on activities in this chapter satisfy the NATEF High Priority-Individual and High Priority-Group requirements. Though there are no NATEF requirements for this section, it is felt that a working knowledge of the industry is necessary.

Tools and equipment needed (NATEF tool list)

- Pen and pencil
- Safety glasses
- Gloves
- Estimating manuals
- Computerized estimating equipment
- Digital camera
- Assorted hand tools

Instructions

The activities for this section are intended to acquaint you with the hand tools. Your instructor will train each of you on the safe use of the tools that you will identify. Safety is a primary concern and you must follow all instructions, both written and demonstrated, and should not operate equipment or use tools without the expressed direction of your instructor.

PROCEDURE

In the lab, after completing the work assignment for that section, follow the work order and record your findings for your instructor to review. If the section calls for "instructor approval," you should not proceed without this approval.

1. After reading the work order, gather the safety gear needed to complete the task. In the space provided below, list the personal and environmental safety equipment and precautions needed for this assignment. Have the instructor check and approve your plan before proceeding.

_____

_____

_____

INSTRUCTOR'S APPROVAL _____

With the vehicle or with the information provided:

2. Write a manual damage report.

INSTRUCTOR COMMENTS:

_____

_____

_____

_____

_____

_____

Name _____ Date _____

Class _____ Instructor _____ Grade _____

1. A damage report is sometimes referred to as a:
   A. police report
   B. damage appraisal
   C. final invoice
   D. phone call that a shop's estimator makes to the owner of a vehicle

2. *Technician A* says that a vehicle cannot be repaired with a manually written damage report. *Technician B* says that a computerized damage report can only be created by the original equipment manufacturer. Who is correct?
   A. Technician A only
   B. Technician B only
   C. Both Technicians A and B
   D. Neither Technician A nor B

3. All of the following can be found on a typical computerized damage report EXCEPT:
   A. the policy owner's deductible
   B. the police report file number
   C. OEM parts number
   D. the vehicle's options

4. **TRUE** or **FALSE:** Sheet metal repair allowances are found in the sheet metal section of a crash manual.

5. *Technician A* says that a Direct Report Program is a repair program that directly involves the insurance company, the repair shop, the original equipment manufacturer, and the agent who sold the policy. *Technician B* says that blending an undamaged panel is also referred to as "fooling the customer." Who is correct?
   A. Technician A only
   B. Technician B only
   C. Both Technicians A and B
   D. Neither Technician A nor B

6. All of the following are important to ascertain while gathering information concerning a collision loss EXCEPT:
   A. the colors of the other vehicles involved in the accident
   B. what temporary repairs were made to the vehicle following the loss
   C. the mileage that the vehicle had at its last inspection
   D. the owner's perception of what might not be working properly since the accident occurred

7. The collision report center's estimator should primarily concern himself or herself with what question while analyzing a collision-damaged vehicle?
   A. Can this vehicle be safely repaired?
   B. How many speeding tickets does the driver have on his or driving record in the past 3 years?
   C. Were any of the drivers involved cited for driving while intoxicated?
   D. Is the vehicle worth less than $10,000?

8. *Technician A* says that a crash guide is published in book form. *Technician B* says that the exploded view of parts that present the most frequent kinds of collision damages are only found in computerized versions of the crash guides. Who is correct?
   A. Technician A only
   B. Technician B only
   C. Both Technicians A and B
   D. Neither Technician A nor B

9. *Technician A* says that 'P-pages' are included in crash manuals only to assist trainee damage estimators in creating a damage report. *Technician B* says digital imaging can be utilized by collision repairers and insurers to correspond with each other. Who is correct?
   A. Technician A only
   B. Technician B only
   C. Both Technicians A and B
   D. Neither Technician A nor B

10. All of the following are typically included in the R&R of a quarter panel EXCEPT:
    A. the antenna
    B. R&I of the rear bumper assembly
    C. dropping of the headliner
    D. the removal and installation of the rear stationary glass

11. Which of the following is NOT important when ordering parts?
    A. included vehicle options
    B. production date of the vehicle
    C. insurance appraiser's code
    D. color and trim codes of the vehicle

12. A work order was created by a repair center for a damaged vehicle. The document—also called repair order—was created from the:
    A. damage report
    B. parts order
    C. VIN
    D. vehicle owner's instructions

13. *Technician A* says that refinish overlap can be applicable on adjacent as well as nonadjacent panels. *Technician B* says that liability coverage is what a first-party collision claim is paid off. Who is correct?
    A. Technician A only
    B. Technician B only
    C. Both Technicians A and B
    D. Neither Technician A nor B

14. *Technician A* says that comprehensive coverage may cover an insured policyholder's collision damage from striking a live deer. *Technician B* says that a deductible is the amount of money deducted from a first-party collision payment. Who is correct?
    A. Technician A only
    B. Technician B only
    C. Both Technicians A and B
    D. Neither Technician A nor B

15. **TRUE** or **FALSE**: Betterment and depreciation charges are collected by the collision repair shop from the owner of the vehicle.

Chapter 11

Collision Damage Analysis

■ WORK ASSIGNMENT 11-1

THE ESSENTIALS OF COLLISION THEORY, IDENTIFYING DIRECT AND INDIRECT DAMAGE, AND VEHICLE DESIGN

Name _____ Date _____

Class _____ Instructor _____ Grade _____

1. After reading the assignment, in the space provided below, list the personal and environmental safety equipment and precautions needed for this assignment.

   _____

   _____

   _____

2. List important information regarding the collision that would be helpful in evaluating the damage; for example: Were the passengers using their seat belts during the collision? Explain why this information is important. Record your findings:

   _____

   _____

   _____

3. What does VIN mean and why is it important? Record your findings:

   _____

   _____

   _____

4. What is a DOT label used for? Record your findings:

   _____

   _____

   _____

5. Describe direct damage and how it can be identified. Record your findings:

   _____

   _____

   _____

6. What is indirect damage and how can it be identified? Record your findings:

   _____

   _____

   _____

7. What is inertia force and how does it contribute to collision damage? Record your findings:

   _____

   _____

8. Describe a BOF design and give examples. Record your findings:

_____

_____

_____

9. What is unibody design and how is it different from BOF? Give examples. Record your findings:

_____

_____

_____

10. Explain how external forces cause damage. Record your findings:

_____

_____

_____

11. Explain how internal forces can affect the following:

Transmission mounts:

_____

_____

Motor mounts:

_____

_____

Shift linkage:

_____

_____

HVAC plenum:

_____

_____

A/C and other lines:

_____

_____

_____

INSTRUCTOR COMMENTS:

_____

_____

_____

_____

_____

_____

■ WORK ASSIGNMENT 11-2

VISUAL INSPECTION AND INSPECTION SEQUENCE

Name _____ Date _____

Class _____ Instructor _____ Grade _____

1. After reading the assignment, in the space provided below, list the personal and environmental safety equipment and precautions needed for this assignment.

_____

_____

_____

2. List a systematic visual inspection sequence and key items that should be checked. Record your findings:

_____

_____

_____

3. What is a vehicle specification manual and how is it helpful? Record your findings:

_____

_____

_____

4. What is a comparative check? Record your findings:

_____

_____

_____

5. How is an X check performed? Record your findings:

_____

_____

_____

6. What is a comparative quick check and why is it important? Record your findings:

_____

_____

_____

7. List the key items that should be checked when inspecting these areas:
   Front fascia/bumper:

_____

_____

   Front lights and electronic sensors:

_____

_____

   Hood and hinges:

_____

_____

_____

Fenders, including badges and trim:

_____

_____

Wheels and tires:

_____

_____

Engine compartment:

_____

_____

Heater core and cooling components:

_____

_____

Doors and hinges:

_____

_____

Roof B pillar and rockers:

_____

_____

Quarter panel:

_____

_____

Rear section:

_____

_____

_____

INSTRUCTOR COMMENTS:

_____

_____

_____

_____

_____

_____

QUICK CHECK MEASUREMENTS AND COLLISION AVOIDANCE EQUIPMENT

Name _____ Date _____

Class _____ Instructor _____ Grade _____

1. After reading the assignment, in the space provided below, list the personal and environmental safety equipment and precautions needed for this assignment.

   _____

   _____

   _____

2. Describe "camber quick check." Record your findings:

   _____

   _____

   _____

3. Explain how a strut "quick check" is performed. Record your findings:

   _____

   _____

   _____

4. Describe how to do a "caster quick check." Record your findings:

   _____

   _____

   _____

5. What is dog tracking? Record your findings:

   _____

   _____

   _____

6. Automobiles are now equipped with very complex and sophisticated equipment that will become damaged in collisions. Please describe briefly the items listed below.

   Adaptive cruise control:

   _____

   _____

   Parking sensors:

   _____

   _____

   Rain sensors:

   _____

   _____

   Lane departure systems:

   _____

   _____

   _____

HUD:

_____

_____

_____

INSTRUCTOR COMMENTS:

_____

_____

_____

_____

_____

_____

THE ESSENTIALS OF COLLISION THEORY, IDENTIFYING DIRECT AND INDIRECT DAMAGE, AND VEHICLE DESIGN

Name _____   Date _____

Class _____   Instructor _____   Grade _____

OBJECTIVES

- Discuss the essentials of collision theory and the effects of force on shape and structural members.
- Discuss the various structural designs used in the manufacture of the automobile and how they are affected by the collision energy.
- Identify and distinguish the difference between direct and indirect damage, including the effects of inertial forces.
- Identify the damage sustained while performing a visual inspection of the damaged vehicle.
- Identify and isolate common and discrete damage sustained by each of the three sections of the automobile.
- Trace the flow of damage caused by a collision.
- Identify the types of quick check measuring methods and techniques that can be used to identify damaged areas of the vehicle.
- Identify the specification manuals and other available resources commonly utilized to identify the damage sustained by the vehicle.

NATEF TASK CORRELATION

The written and hands-on activities in this chapter satisfy the NATEF High Priority-Individual and High Priority-Group requirements for Section I: A. 1, 2, 3, 4, 5, 6, 7, 10, 17; B. 1, 7, 21.

Tools and equipment needed (NATEF tool list)

- Safety glasses
- Gloves
- Ear protection
- Particle mask
- Pencil and paper
- Assorted hand tools

Vehicle Description

Year_____ Make _____ Model _____

VIN _____ Paint Code _____

PROCEDURE

1. Gather the safety gear needed to complete the task. In the space provided below, list the personal and environmental safety equipment needed and the precautions necessary for this assignment. Have the instructor check and approve your plan before proceeding.

_____

_____

_____

INSTRUCTOR'S APPROVAL _____

2. Describe the type of construction of the vehicle. Record your findings:

_____

_____

_____

3. Retrieve the VIN and decode the country of origin. Record your findings:

_____

4. What is the date of manufacture? Record your findings:

_____

5. Retrieve the paint code and convert it to your company code. Record your findings:

_____

6. Identify the direct damage to the vehicle and, using the correct terms, list the corrective measures that are necessary. Record your findings:

_____

_____

_____

7. Identify the indirect damage to the vehicle and, using the correct terms, list the corrective measures that are necessary. Record your findings:

_____

_____

_____

8. Identify the inertial damage to the vehicle and, using the correct terms, list the corrective measures that are necessary. Record your findings:

_____

_____

_____

INSTRUCTOR COMMENTS:

_____

_____

_____

_____

_____

_____

VISUAL INSPECTION, ISOLATING INTO SECTIONS, AND INERTIAL DAMAGE

Name _____ Date _____

Class _____ Instructor _____ Grade _____

OBJECTIVES

- Discuss the essentials of collision theory and the effects of force on shape and structural members.
- Discuss the various structural designs used in the manufacture of the automobile and how they are affected by the collision energy.
- Identify and distinguish the difference between direct and indirect damage, including the effects of inertial forces.
- Identify the damage sustained while performing a visual inspection of the damaged vehicle.
- Identify and isolate common and discrete damage sustained by each of the three sections of the automobile.
- Trace the flow of damage caused by a collision.
- Identify the types of quick check measuring methods and techniques that can be used to identify damaged areas of the vehicle.
- Identify the specification manuals and other available resources commonly utilized to identify the damage sustained by the vehicle.

NATEF TASK CORRELATION

The written and hands-on activities in this chapter satisfy the NATEF High Priority-Individual and High Priority-Group requirements for Section I: A. 1, 2, 3, 4, 5, 6, 7, 10, 17; B. 1, 7, 21.

Tools and equipment needed (NATEF tool list)

- Safety glasses
- Gloves
- Ear protection
- Particle mask
- Pencil and paper
- Assorted hand tools

Vehicle Description

Year_____ Make _____ Model _____

VIN _____ Paint Code _____

PROCEDURE

1. Gather the safety gear needed to complete the task. In the space provided below, list the personal and environmental safety equipment needed and the precautions necessary for this assignment. Have the instructor check and approve your plan before proceeding.

_____

_____

_____

INSTRUCTOR'S APPROVAL _____

2. Using a vehicle specification manual, manually write a damage report on a minor collision vehicle using visual inspection only. You will need to obtain a blank damage estimate sheet from your instructor.

INSTRUCTOR COMMENTS:

_____

_____

_____

_____

_____

_____

QUICK CHECK MEASUREMENTS AND USING REPAIR INFORMATION GUIDES

Name _____ Date _____

Class _____ Instructor _____ Grade _____

OBJECTIVES

- Discuss the essentials of collision theory and the effects of force on shape and structural members.
- Discuss the various structural designs used in the manufacture of the automobile and how they are affected by the collision energy
- Identify and distinguish the difference between direct and indirect damage, including the effects of inertial forces.
- Identify the damage sustained while performing a visual inspection of the damaged vehicle.
- Identify and isolate common and discrete damage sustained by each of the three sections of the automobile.
- Trace the flow of damage caused by a collision.
- Identify the types of quick check measuring methods and techniques that can be used to identify damaged areas of the vehicle.
- Identify the specification manuals and other available resources commonly utilized to identify the damage sustained by the vehicle.

NATEF TASK CORRELATION

The written and hands-on activities in this chapter satisfy the NATEF High Priority-Individual and High Priority-Group requirements for Section I: A. 1, 2, 3, 4, 5, 6, 7, 10, 17; B 1, 7, 21.

Tools and equipment needed (NATEF tool list)

- Safety glasses
- Gloves
- Ear protection
- Particle mask
- Pencil and paper
- Assorted hand tools

Vehicle Description

Year_____ Make _____ Model _____

VIN _____ Paint Code _____

PROCEDURE

1. Gather the safety gear needed to complete the task. In the space provided below, list the personal and environmental safety equipment needed and the precautions necessary for this assignment. Have the instructor check and approve your plan before proceeding.

_____

_____

_____

INSTRUCTOR'S APPROVAL _____

2. Using a vehicle specification manual, manually or by a computerized estimating system, write a damage report on a minor collision vehicle using a thorough systematic approach. You will need to obtain a blank damage estimate sheet from your instructor.

INSTRUCTOR COMMENTS:

_____

_____

_____

_____

_____

_____

Name _____ Date _____

Class _____ Instructor _____ Grade _____

1. *Technician A* says that the inertial forces frequently cause the damage immediately around the point of impact. *Technician B* says that the point of impact and the direct damage are usually found in the same area of the vehicle. Who is correct?
   A. Technician A only
   B. Technician B only
   C. Both Technicians A and B
   D. Neither Technician A nor B

2. *Technician A* says that many vehicle manufacturers use sacrificial structures to deflect the effects of some of the impact forces. *Technician B* says that mass weight movement can cause damage to the engine and transmission mounts. Who is correct?
   A. Technician A only
   B. Technician B only
   C. Both Technicians A and B
   D. Neither Technician A nor B

3. *Technician A* says that hybrid vehicles are either a mild hybrid or a full hybrid. *Technician B* says that aluminum parts lack the flexibility of steel and become work hardened more easily than steel. Who is correct?
   A. Technician A only
   B. Technician B only
   C. Both Technicians A and B
   D. Neither Technician A nor B

4. *Technician A* says that many of the sensitive warning and informational sensors are placed at the front of the vehicle in the area of the grille and radiator support. *Technician B* says that some vehicles have an inertial shutoff switch that automatically turns the ignition off under certain operating conditions. Who is correct?
   A. Technician A only
   B. Technician B only
   C. Both Technicians A and B
   D. Neither Technician A nor B

5. *Technician A* says that the VIN is a 17-digit number that provides vehicle-specific information to the technician. *Technician B* says that the antitheft labels have the same number on them as does the VIN. Who is correct?
   A. Technician A only
   B. Technician B only
   C. Both Technicians A and B
   D. Neither Technician A nor B

6. Which of the following would NOT normally be used to make comparative measures on the undercarriage when checking for damage?
   A. tram bar
   B. self-centering gauges
   C. tape measure

7. *Technician A* says that crush zones are built into the A pillar for added reinforcement. *Technician B* says that reinforcement foam need not be replaced if some of it comes off when removing the damaged panel. Who is correct?
   A. Technician A only
   B. Technician B only
   C. Both Technicians A and B
   D. Neither Technician A nor B

8. *Technician A* says that the collision damage radiates throughout the vehicle many times, damaging panels a considerable distance from the point of impact. *Technician B* says that the inertial forces are frequently the cause of the damage that occurs in a part of the vehicle away from the impact area. Who is correct?
   A. Technician A only
   B. Technician B only
   C. Both Technicians A and B
   D. Neither Technician A nor B

9. *Technician A* says that knowing the speed of the vehicle and the angle of impact is helpful to make a more accurate damage assessment. *Technician B* says that using a recycled part may reduce some of the corrosion problems. Who is correct?
   A. Technician A only
   B. Technician B only
   C. Both Technicians A and B
   D. Neither Technician A nor B

10. *Technician A* says that a vehicle skidding sideways during a collision can affect the front suspension parts. *Technician B* says that the damage sustained by a body over frame vehicle tends to remain more localized than on a unibody. Who is correct?
    A. Technician A only
    B. Technician B only
    C. Both Technicians A and B
    D. Neither Technician A nor B

11. *Technician A* says that crush zones are built into the A pillar for added reinforcement. *Technician B* says that reinforcement foam need not be replaced if some of it comes off when removing the damaged panel. Who is correct?
    A. Technician A only
    B. Technician B only
    C. Both Technicians A and B
    D. Neither Technician A nor B

Chapter 12

Bolted Exterior Panel Replacement

■ WORK ASSIGNMENT 12-1

FRONT-END PART REPLACEMENT

Name _____   Date _____

Class _____   Instructor _____   Grade _____

NATEF TASKS I C. 1, 7, 2; II B. 1, 2, 3, 4, 5, 6, 7, 8, 11, 12; D. 1, 2, 3, 4

1. After reading the assignment, in the space provided below, list the personal and environmental safety equipment and precautions needed for this assignment.

   _____

   _____

   _____

2. List the common fasteners other than bolts that are used for vehicle construction. Record your findings:

   _____

   _____

3. Why is it important to record the fasteners used for each part? Record your findings:

   _____

   _____

4. Why do plastic and composite fenders require different bolts? Record your findings:

   _____

   _____

5. How would a technician adjust a fender that is too tight to the door, causing rubbing when it is opened? Record your findings:

   _____

   _____

   _____

6. Why is proper adjustment of a core support so important? Record your findings:

   _____

   _____

   _____

7. What are shims used for when adjusting a fender? Record your findings:

_____

_____

_____

8. When installing a hood the bolts are placed through the hinge, then the hood is carefully closed to check alignment. If the hood gap is too large on the left rear corner and too tight on the right front, what can be done to correct the problem? Record your findings:

_____

_____

_____

INSTRUCTOR COMMENTS:

_____

_____

_____

_____

_____

_____

_____

■ WORK ASSIGNMENT 12-2

BUMPER, FASCIA, AND DOORS

Name _____ Date _____

Class _____ Instructor _____ Grade _____

NATEF TASKS I C. 1, 7, 2; II B. 1, 2, 3, 4, 5, 6, 7, 8, 11, 12; D. 1, 2, 3, 4

1. After reading the assignment, in the space provided below, list the personal and environmental safety equipment and precautions needed for this assignment.

   _____

   _____

   _____

2. What is the difference between a steel bumper and a fascia? Record your findings:

   _____

   _____

   _____

3. What is a face bar? Record your findings:

   _____

   _____

   _____

4. Describe a head light and bezel. Record your findings:

   _____

   _____

   _____

5. Who does a bolted door hinge adjustment slot work? Record your findings:

   _____

   _____

   _____

6. List the steps needed to remove a door. Record your findings:

   _____

   _____

   _____

7. How is a reinstalled door adjusted? Record your findings:

   _____

   _____

   _____

8. How is a door adjustment fine-tuned? Record your findings:

   _____

   _____

   _____

9. What is a door striker and how and when should it be installed? Record your findings:

_____

_____

_____

INSTRUCTOR COMMENTS:

_____

_____

_____

_____

_____

_____

DECK LID, HATCH, REAR LIGHTS, PICKUP BOX, AND GLASS

Name _____   Date _____

Class _____   Instructor _____   Grade _____

NATEF TASKS I C. 1, 7, 2; II B. 1, 2, 3, 4, 5, 6, 7, 8, 11, 12; D. 1, 2, 3, 4

1. After reading the assignment, in the space provided below, list the personal and environmental safety equipment and precautions needed for this assignment.

   _____

   _____

   _____

   _____

2. How is a deck lid removed? Record your findings:

   _____

   _____

   _____

3. How is a deck lid adjusted? Record your findings:

   _____

   _____

   _____

4. Describe a torsion bar-type deck hinge. Record your findings:

   _____

   _____

   _____

5. How does removing and installing a composite part differ from steel? Record your findings:

   _____

   _____

   _____

6. Describe how to remove a pickup box. Record your findings:

   _____

   _____

   _____

7. What is LSG and how does it differ from tempered glass? Record your findings:

   _____

   _____

   _____

8. What is a glass regulator? Record your findings:

   _____

   _____

   _____

INSTRUCTOR COMMENTS:

■ WORK ORDER 12-1

FRONT-END PART REPLACEMENT

Name _____ Date _____

Class _____ Instructor _____ Grade _____

OBJECTIVES

- Discuss the purpose of attaching parts using bolts or other types of fasteners commonly used to secure exterior parts.
- Discuss the different types of materials used for manufacturing the component parts.
- Discuss the structural function that the component parts offer.
- Describe the attaching methods used and the basic concepts for installing and adjusting the techniques used.
- Explain the types of fasteners commonly used on these panels.

NATEF TASK CORRELATION

The written and hands-on activities in this chapter satisfy the NATEF High Priority-Individual and High Priority-Group requirements for Section NATEF Tasks I C. 1, 7, 2; II B. 1, 2, 3, 4, 5, 6, 7, 8, 11, 12; D. 1, 2, 3, 4.

Tools and equipment needed (NATEF tool list)

- Safety glasses
- Ear protection
- Pencil and paper
- Vehicle
- Vehicle service manual
- Plastic bags
- Gloves
- Particle mask
- Assorted hand tools
- Estimate or work order for the vehicle
- Assorted shims and fasteners
- Storage cart

Vehicle Description

Year_____ Make _____ Model _____

VIN _____ Paint Code _____

PROCEDURE

1. After reading the assignment, in the space provided below, list the personal and environmental safety equipment and precautions needed for this assignment.

_____

_____

_____

INSTRUCTOR'S APPROVAL _____

Using the vehicle and work order/estimate provided by your instructor, complete the task below.

2. Remove the front fender of the vehicle. Record your sequence:

    _____

    _____

    _____

3. Bag and tag the fasteners that were removed.
    Instructor comment:

    _____

    _____

    _____

4. Reinstall and adjust the fender.
    Instructor comment:

    _____

    _____

    _____

5. Remove the hood from the vehicle. Record your sequence:

    _____

    _____

    _____

6. Bag and tag the fasteners that were removed.

7. Remove the hood hinges from the vehicle. Record your sequence:

    _____

    _____

    _____

8. Bag and tag the fasteners that were removed.

9. Remove the hood latch from the vehicle. Record your sequence:

    _____

    _____

    _____

10. Bag and tag the fasteners that were removed.

11. Reinstall and adjust the hood, hood hinges, and latch.

INSTRUCTOR COMMENTS:

_____

_____

_____

_____

_____

_____

BUMPER, FASCIA, AND DOORS

Name _____ Date _____

Class _____ Instructor _____ Grade _____

OBJECTIVES

- Discuss the purpose of attaching parts using bolts or other types of fasteners commonly used to secure exterior parts.
- Discuss the different types of materials used for manufacturing the component parts.
- Discuss the structural function that the component parts offer.
- Describe the attaching methods used and the basic concepts for installing and adjusting the techniques used.
- Explain the types of fasteners commonly used on these panels.

NATEF TASK CORRELATION

The written and hands-on activities in this chapter satisfy the NATEF High Priority-Individual and High Priority-Group requirements for NATEF Tasks I C. 1, 7, 2; II B. 1, 2, 3, 4, 5, 6, 7, 8, 11, 12; D. 1, 2, 3, 4.

Tools and equipment needed (NATEF tool list)

- Safety glasses
- Ear protection
- Pencil and paper
- Vehicle
- Vehicle service manual
- Plastic bags

- Gloves
- Particle mask
- Assorted hand tools
- Estimate or work order for the vehicle
- Assorted shims and fasteners
- Storage cart

Vehicle Description

Year_____ Make _____ Model _____

VIN _____ Paint Code _____

PROCEDURE

1. After reading the assignment, in the space provided below, list the personal and environmental safety equipment and precautions needed for this assignment.

_____

_____

_____

INSTRUCTOR'S APPROVAL _____

Using the vehicle and work order/estimate provided by your instructor, complete the task below.

2. Remove the front steel bumper of the vehicle. Record your sequence:

_____

_____

_____

3. Bag and tag the fasteners that were removed.
   Instructor comment:

   _____

   _____

   _____

4. Reinstall and adjust the bumper.
   Instructor comment:

   _____

   _____

   _____

5. Remove the fascia of the vehicle. Record your sequence:

   _____

   _____

   _____

6. Bag and tag the fasteners that were removed.
   Instructor comment:

   _____

   _____

   _____

7. Reinstall and adjust the fascia.
   Instructor comment:

   _____

   _____

   _____

8. Remove the door from the vehicle. Record your sequence:

   _____

   _____

   _____

9. Bag and tag the fasteners that were removed.
10. Remove the door hinges from the vehicle. Record your sequence:

    _____

    _____

    _____

11. Bag and tag the fasteners that were removed.
12. Remove the door striker from the vehicle. Record your sequence:

    _____

    _____

    _____

13. Bag and tag the fasteners that were removed.
    Instructor comment:

    _____

    _____

    _____

14. Reinstall and adjust the door, door hinges, and striker.

INSTRUCTOR COMMENTS:

_____

_____

_____

_____

_____

_____

DECK LID, HATCH, REAR LIGHTS, PICKUP BOX, AND GLASS

Name _____   Date _____

Class _____   Instructor _____   Grade _____

OBJECTIVES

- Discuss the purpose of attaching parts using bolts or other types of fasteners commonly used to secure exterior parts.
- Discuss the different types of materials used for manufacturing the component parts.
- Discuss the structural function that the component parts offer.
- Describe the attaching methods used and the basic concepts for installing and adjusting the techniques used.
- Explain the types of fasteners commonly used on these panels.

NATEF TASK CORRELATION

The written and hands-on activities in this chapter satisfy the NATEF High Priority-Individual and High Priority-Group requirements for NATEF Tasks I C. 1, 7, 2; II B. 1, 2, 3, 4, 5, 6, 7, 8, 11, 12; D. 1, 2, 3, 4.

Tools and equipment needed (NATEF tool list)

- Safety glasses
- Ear protection
- Pencil and paper
- Vehicle
- Vehicle service manual
- Plastic bags
- Gloves
- Particle mask
- Assorted hand tools
- Estimate or work order for the vehicle
- Assorted shims and fasteners
- Storage cart

Vehicle Description

Year_____ Make _____ Model _____

VIN _____ Paint Code _____

PROCEDURE

1. After reading the assignment, in the space provided below, list the personal and environmental safety equipment and precautions needed for this assignment.

_____

_____

_____

INSTRUCTOR'S APPROVAL _____

Using the vehicle and work order/estimate provided by your instructor, complete the task below.

2. Remove the rear deck from the vehicle. Record your sequence:

_____

_____

_____

3. Bag and tag the fasteners that were removed.
   Instructor comment:

_____

_____

_____

4. Reinstall and adjust the deck.
   Instructor comment:

_____

_____

_____

5. Remove the hatch from the vehicle. Record your sequence:

_____

_____

_____

6. Bag and tag the fasteners that were removed.
   Instructor comment:

_____

_____

_____

7. Reinstall and adjust the hatch.
   Instructor comment:

_____

_____

_____

8. Remove the back lights from the vehicle. Record your sequence:

_____

_____

_____

9. Bag and tag the fasteners that were removed.
   Instructor comment:

_____

_____

_____

10. Reinstall and adjust the back lights.
    Instructor comment:

_____

_____

_____

11. Remove the box from the pickup. Record your sequence:

_____

12. Bag and tag the fasteners that were removed.
    Instructor comment:

    _____

    _____

    _____

13. Reinstall and adjust the box.
    Instructor comment:

    _____

    _____

    _____

14. Remove the front door glass from the vehicle. Record your sequence:

    _____

    _____

    _____

15. Bag and tag the fasteners that were removed.
    Instructor comment:

    _____

    _____

    _____

16. Reinstall and adjust the door glass.
    Instructor comment:

    _____

    _____

    _____

17. Remove the front steel bumper of the vehicle. Record your sequence:

    _____

    _____

    _____

18. Bag and tag the fasteners that were removed.
    Instructor comment:

    _____

    _____

    _____

19. Reinstall and adjust the bumper.

INSTRUCTOR COMMENTS:

_____

_____

_____

_____

_____

_____

Name _____ Date _____

Class _____ Instructor _____ Grade _____

1. *Technician A* says that the air bag sensors for the side curtains are located at the front of the vehicle next to the front air bag sensors. *Technician B* says that the air bags should be disabled before attempting to service the door if the vehicle is equipped with side air bags. Who is correct?
   A. Technician A only
   B. Technician B only
   C. Both Technicians A and B
   D. Neither Technician A nor B

2. *Technician A* says that manufacturers use component panels to make access to mechanical parts easier. *Technician B* says that bolts and nuts and clips are the most common methods used to secure component parts. Who is correct?
   A. Technician A only
   B. Technician B only
   C. Both Technicians A and B
   D. Neither Technician A nor B

3. *Technician A* says that some fasteners are one-time use and must be replaced when parts are replaced. *Technician B* says that composite panels may require a different style of hardware for securing them than does steel. Who is correct?
   A. Technician A only
   B. Technician B only
   C. Both Technicians A and B
   D. Neither Technician A nor B

4. *Technician A* says that adjusting front sheet metal is more restrictive on the unibody vehicle than on the BOF vehicle. *Technician B* says one may X check the front end to determine if it is out of proper alignment. Who is correct?
   A. Technician A only
   B. Technician B only
   C. Both Technicians A and B
   D. Neither Technician A nor B

5. *Technician A* says that the hood has two catches, the primary and secondary, for emergencies. *Technician B* says that the hood latch should be the first thing to attach when adjusting the hood. Who is correct?
   A. Technician A only
   B. Technician B only
   C. Both Technicians A and B
   D. Neither Technician A nor B

6. *Technician A* says that adjusting doors should be done by loosening all bolts on both hinges and moving the door. *Technician B* says that the fore and aft adjustment is made on the hinge-to-pillar location. Who is correct?
   A. Technician A only
   B. Technician B only
   C. Both Technicians A and B
   D. Neither Technician A nor B

7. *Technician A* says that the hatch on a hatchback model vehicle is attached to the rear of the roof with hinges. *Technician B* says that LSG is typically used for windshields. Who is correct?
   A. Technician A only
   B. Technician B only
   C. Both Technicians A and B
   D. Neither Technician A nor B

8. *Technician A* says that one must torque the fasteners in a specific sequence on composite panels. *Technician B* says that electrical plungers are sometimes used in place of hard electric wires. Who is correct?
   A. Technician A only
   B. Technician B only
   C. Both Technicians A and B
   D. Neither Technician A nor B

9. *Technician A* says that the bumper usually has either reinforcement or an energy absorber. *Technician B* says that the reinforcement is made of low carbon to bend so it can absorb some of the impact forces. Who is correct?
    A. Technician A only
    B. Technician B only
    C. Both Technicians A and B
    D. Neither Technician A nor B

10. *Technician A* says that one should start to torque the fenders down first at the front and work back toward the door. *Technician B* says that a positive flushness may be required between the rear of the front fender and the door. Who is correct?
    A. Technician A only
    B. Technician B only
    C. Both Technicians A and B
    D. Neither Technician A nor B

11. Which of the following statements is correct?
    A. On a body over frame vehicle the front panels are attached to the frame via an adjustable radiator core support.
    B. The front fenders should be shimmed to raise them up to the correct height to compensate for insufficient structural repair.
    C. The hood should be installed last when replacing the entire front sheet metal assembly.
    D. The rear glass is nearly always attached to the rear hatch with hinges on a hatchback model.

12. *Technician A* says that tempered glass is used on nearly all side windows of a car. *Technician B* says that the rubber bumpers on top of the core support are used to eliminate the flutter on the hood. Who is correct?
    A. Technician A only
    B. Technician B only
    C. Both Technicians A and B
    D. Neither Technician A nor B

13. *Technician A* says that an egg crate design, high-density foam, and a collapsible mechanical piston are all designs used for energy absorbers. *Technician B* says that thermal expansion occurs on all exterior panels. Who is correct?
    A. Technician A only
    B. Technician B only
    C. Both Technicians A and B
    D. Neither Technician A nor B

14. *Technician A* says that the hinges on some vehicles are welded to both the door and the hinge pillar. *Technician B* says that even though they are welded to the body, the door hinges can still be adjusted. Who is correct?
    A. Technician A only
    B. Technician B only
    C. Both Technicians A and B
    D. Neither Technician A nor B

15. *Technician A* says that the radiator support is attached to the frame rail with a large bushing or isolator between the two. *Technician B* says that the radiator support can be adjusted ½ inch in either direction to accommodate properly aligning the panels. Who is correct?
    A. Technician A only
    B. Technician B only
    C. Both Technicians A and B
    D. Neither Technician A nor B

Chapter 13

Measuring Structural Damage

■ WORK ASSIGNMENT 13-1

SAFETY, VEHICLE INSPECTION, AND REPAIR PLAN

Name _____   Date _____

Class _____   Instructor _____   Grade _____
$

NATEF TASK I A. 1, 10, 11, 14, 15, 16; B. 1, 3, 4, 5, 6, 7, 21

1. After reading the assignment, in the space provided below, list the personal and environmental safety precautions necessary for this assignment.

   _____

   _____

   _____

2. Though safety is important in all areas of collision repair, list some reasons why safety should be of a special concern to a technician when working with frame equipment. Record your findings:

   _____

   _____

   _____

3. Why is making a repair plan and measuring the vehicle so important when preparing to repair a vehicle? Record your findings:

   _____

   _____

   _____

4. How are primary damage and POI related? Record your findings:

   _____

   _____

   _____

5. How are misalignment and secondary damage related? Record your findings:

   _____

   _____

   _____

6. Why is a walkaround inspection before measuring important to the measuring process? Record your findings:

   _____

   _____

   _____

7. What areas of the vehicle may have damage away from the POI following a front-end damage collision? Explain. Record your findings:

_____

_____

_____

8. Why does a rear damaged collision react differently from a front-end collision? Record your findings:

_____

_____

_____

9. Why or how does a side impact damaged vehicle shorten its length? Record your findings:

_____

_____

_____

10. What part of the vehicle will be affected by a rollover collision? Record your findings:

_____

_____

_____

11. How does inertia affect a vehicle in a collision? Record your findings:

_____

_____

_____

INSTRUCTOR COMMENTS:

_____

_____

_____

_____

_____

_____

MEASURING TECHNIQUES

Name _____   Date _____

Class _____   Instructor _____   Grade _____

NATEF TASKS I A. 1, 10, 11, 14, 15, 16; B. 1, 3, 4, 5, 6, 7, 21

1. After reading the assignment, in the space provided below, list the personal and environmental safety precautions necessary for this assignment.

   _____

   _____

   _____

2. Why is the center section of the vehicle important when analyzing the condition of a vehicle following a collision? Record your findings:

   _____

   _____

   _____

3. Explain the types of damage listed below.

   Sideway:

   _____

   _____

   _____

   Sag:

   _____

   _____

   _____

   Mash:

   _____

   _____

   _____

   Twist:

   _____

   _____

   _____

   Diamond:

   _____

   _____

   _____

4. Explain the types of measuring below.

   Point-to-point:

   _____

   _____

   _____

Tape measuring:

_____

_____

_____

Checking for square:

_____

_____

_____

Tram bar/gauge:

_____

_____

_____

5. How do measuring irregular shapes differ? Record your findings:

_____

_____

_____

INSTRUCTOR COMMENTS:

_____

_____

_____

_____

_____

_____

■ WORK ASSIGNMENT 13-3

ZERO PLANE, CENTERLINE, AND DATUM PLANE

Name _____   Date _____

Class _____   Instructor _____   Grade _____

NATEF TASKS I A. 1, 10, 11, 14, 15, 16; B 1, 3, 4, 5, 6, 7, 21

1. After reading the assignment, in the space provided below, list the personal and environmental and safety precautions necessary for this assignment.

   _____

   _____

   _____

2. What is a torque box and how is it related to the zero plane? Record your findings:

   _____

   _____

   _____

3. How many undamaged reference points are necessary to begin accurate measurement? Record your findings:

   _____

   _____

   _____

4. Are datum measurements and point-to-point measurements the same? If not, how do they differ? Record your findings:

   _____

   _____

   _____

5. What is a datum line measurement? Record your findings:

   _____

   _____

   _____

6. What is the centerline of a vehicle and how is it used to measure a vehicle? Record your findings:

   _____

   _____

   _____

7. What is a comparative measurement? Record your findings:

   _____

   _____

   _____

INSTRUCTOR COMMENTS:

_____

_____

_____

_____

_____

THREE-DIMENSIONAL MEASURING SYSTEMS

Name _____ Date _____

Class _____ Instructor _____ Grade _____

NATEF TASKS I A. 1, 10, 11, 14, 15, 16; B. 1, 3, 4, 5, 6, 7, 21

1. After reading the assignment, in the space provided below, list the personal and environmental safety precautions necessary for this assignment.

_____

_____

_____

2. What does a three-dimensional measuring system measure and why is it important? Record your findings:

_____

_____

_____

3. Briefly describe the measuring systems below.

Centering gauges:

_____

_____

_____

How are they used and read?

_____

_____

_____

Strut tower gauges:

_____

_____

_____

Dedicated measuring system:

_____

_____

_____

Universal measuring system:

_____

_____

_____

Fixture measuring system:

_____

_____

_____

Computerized measuring system:

_____

_____

_____

Laser measuring system:

_____

_____

_____

Ultrasonic measuring system:

_____

_____

_____

INSTRUCTOR COMMENTS:

_____

_____

_____

_____

_____

_____

1. *Technician A* says that the point zero plane is the area from which most of the undercarriage measurements are taken. *Technician B* says that diagonal measurements should extend from one section of the vehicle into the next. Who is correct?
   A. Technician A only
   B. Technician B only
   C. Both Technicians A and B
   D. Neither Technician A nor B

2. Which of the following is NOT considered part of the three-dimensional measuring process?
   A. length
   B. width
   C. height
   D. depth

3. Which of the following is NOT a computerized measuring system?
   A. laser system
   B. jig and fixture system
   C. electromechanical system
   D. sonic measuring system

4. *Technician A* says that the centering gauges can be used to determine the datum plane. *Technician B* says that the height measurements are taken from the datum plane. Who is correct?
   A. Technician A only
   B. Technician B only
   C. Both Technicians A and B
   D. Neither Technician A nor B

5. *Technician A* says that the same areas used as the control points for repair purposes are used by the manufacturer as holding fixture locations. *Technician B* says that the center plane is used to take side-to-side measurements. Who is correct?
   A. Technician A only
   B. Technician B only
   C. Both Technicians A and B
   D. Neither Technician A nor B

6. *Technician A* says that the jig and fixture system is usually a "go or no go" measuring system. *Technician B* says that the universal mechanical measuring system frequently requires removing several suspension parts prior to securing it to the bench. Who is correct?
   A. Technician A only
   B. Technician B only
   C. Both Technicians A and B
   D. Neither Technician A nor B

7. *Technician A* says that the damage will radiate deeper into the unibody on a front hit than it will on a rear impact with a comparable force. *Technician B* says that in a collision, each section of the unibody responds in a like manner to the section in front or behind it. Who is correct?
   A. Technician A only
   B. Technician B only
   C. Both Technicians A and B
   D. Neither Technician A nor B

8. *Technician A* says that when making point-to-point measurements with a tram gauge, the pins should always be adjusted so the bar can be held parallel to the datum plane. *Technician B* says that one long diagonal measurement is more accurate than three shorter ones. Who is correct?
   A. Technician A only
   B. Technician B only
   C. Both Technicians A and B
   D. Neither Technician A nor B

9. *Technician A* says that most of the indirect damage is caused by the inertial forces in a more severe collision. *Technician B* says that most three-dimensional measuring systems require two correct dimensional locations to establish the system. Who is correct?
   A. Technician A only
   B. Technician B only
   C. Both Technicians A and B
   D. Neither Technician A nor B

10. Which of the following would be the most usable for taking point-to-point measurements on the undercarriage?
    A. tram bar
    B. self-centering gauge
    C. tape measure
    D. strut tower gauge

11. *Technician A* says that the electromechanical measuring system is able to monitor multiple locations while the repairs are being made. *Technician B* says that most of the computerized measuring systems are dedicated to only one specific pulling or straightening system. Who is correct?
    A. Technician A only
    B. Technician B only
    C. Both Technicians A and B
    D. Neither Technician A nor B

12. *Technician A* says that a diamond damage condition exists when the frame rail on one side of the vehicle has been raised at the front and lowered at the rear in a collision. *Technician B* says that the dimensions from the center to the outside of the vehicle are equal on both sides on an asymmetrical vehicle. Who is correct?
    A. Technician A only
    B. Technician B only
    C. Both Technicians A and B
    D. Neither Technician A nor B

13. *Technician A* says that the roof of a vehicle that is struck hard in the center of the side will lean toward the side of the vehicle that was struck. *Technician B* says that because the indirect damage occurred last, it should not be removed until after the direct damage is roughed out. Who is correct?
    A. Technician A only
    B. Technician B only
    C. Both Technicians A and B
    D. Neither Technician A nor B

14. *Technician A* says that when making point-to-point measurements where a large hole is used as a reference point, it may be advisable to measure edge to edge for more accuracy. *Technician B* says that all the damaged exterior parts should be removed before anything else is done to gain access to any areas where damage may exist. Who is correct?
    A. Technician A only
    B. Technician B only
    C. Both Technicians A and B
    D. Neither Technician A nor B

15. *Technician A* says that most measuring reference points are taken to the center of the bolts, bolt heads, holes, and rivets. *Technician B* says that occasionally it may be necessary to make comparative side-to-side measurements at reference points that are not used in the undercarriage specification manual. Who is correct?
    A. Technician A only
    B. Technician B only
    C. Both Technicians A and B
    D. Neither Technician A nor B

Chapter 14

Straightening and Repairing Structural Damage

■ WORK ASSIGNMENT 14-1

ANALYZING COLLISION DAMAGE

Name _____   Date _____

Class _____   Instructor _____   Grade _____

NATEF TASKS A. 2, 3, 4. 5, 6, 7, 8, 10, 11, 12, 13, 15, 17; B. 1, 2, 4, 5, 6, 7, 8, 9, 10, 11, 12, 13, 14, 15, 16, 17, 18, 19, 20, 21

1. After reading the assignment, in the space provided below, list the personal and environmental safety equipment and precautions needed for this assignment.

   _____
   _____
   _____

2. What is a hostile force in a collision, and how does it affect the vehicle that is involved in a collision? Record your findings:

   _____
   _____
   _____

3. Why is identifying the path of damage important, and how does its effect change as it travels deeper into the vehicle? Record your findings:

   _____
   _____
   _____

4. List some "tell-tale" signs of damage. Record your findings:

   _____
   _____
   _____

5. List some of the subtle signs of damage that should be looked for when analyzing a vehicle. Record your findings:

   _____
   _____
   _____

6. Describe direct damage and how it affects a vehicle in a collision. Record your findings:

   _____
   _____

7. Describe indirect damage and how it affects a vehicle in a collision. Record your findings:

_____

_____

_____

8. Why would it be necessary to disassemble a vehicle to analyze the damage? Record your findings:

_____

_____

_____

INSTRUCTOR COMMENTS:

_____

_____

_____

_____

_____

_____

STRAIGHTENING EQUIPMENT AND DAMAGE CONCEPTS

Name _____   Date _____

Class _____   Instructor _____   Grade _____

NATEF TASKS A. 2, 3, 4. 5, 6, 7, 8, 10, 11, 12, 13, 15, 17; B. 1, 2, 4, 5, 6, 7, 8, 9, 10, 11, 12, 13, 14, 15, 16, 17, 18, 19, 20, 21

1. After reading the assignment, in the space provided below, list the personal and environmental safety equipment and precautions needed for this assignment.

   _____

   _____

   _____

2. Safety is of extreme importance when working with straightening equipment. List some of the specific precautions that a technician should take when working with this type of equipment. Record your reply:

   _____

   _____

3. Briefly describe a floor-mounted system. Record your findings:

   _____

   _____

4. What is a dedicated fixture system, and how does it differ from other straightening equipment? Record your findings:

   _____

   _____

5. Describe a "rack system" and its advantages. Record your findings:

   _____

   _____

6. Why is anchoring important? Record your findings:

   _____

   _____

7. Describe how to anchor to a pinch weld. Record your findings:

   _____

   _____

8. Describe how to anchor a BOF vehicle. Record your findings:

   _____

   _____

   _____

9. Why is it important to choose a pulling location correctly? Record your findings:

_____

_____

_____

10. What is stress relieving? Record your findings:

_____

_____

_____

11. How does heat affect steel, and what are some of the precautions a technician should follow? Record your findings:

_____

_____

_____

12. Describe how inertia acts in a collision. Record your findings:

_____

_____

_____

INSTRUCTOR COMMENTS:

_____

_____

_____

_____

_____

_____

_____

COMMON DAMAGE AND REPAIR PROCEDURES

Name _____   Date _____

Class _____   Instructor _____   Grade _____

NATEF TASKS A. 2, 3, 4. 5, 6, 7, 8, 10, 11, 12, 13, 15, 17; B. 1, 2, 4, 5, 6, 7, 8, 9, 10, 11, 12, 13, 14, 15, 16, 17, 18, 19, 20, 21

1. After reading the assignment, in the space provided below, list the personal and environmental safety equipment and precautions needed for this assignment.

   _____

   _____

   _____

2. What types and location of damage should a technician look for in a rear collision? Record your findings:

   _____

   _____

   _____

3. What is a rough pull? Record your findings:

   _____

   _____

   _____

4. What does laminated steel do for the vehicle, and how should it be repaired? Record your findings:

   _____

   _____

   _____

5. What makes a side impact collision difficult, and how is the vehicle's correct length restored? Record your findings:

   _____

   _____

   _____

6. What makes a roll-over collision difficult to analyze and repair? Record your findings:

   _____

   _____

   _____

7. Describe the damage conditions listed below and how to identify them.
   Diamond:

   _____

   _____

   _____

   Twist:

   _____

   _____

   _____

Mash:

_____

_____

_____

Side sway:

_____

_____

_____

Sag/kick up:

_____

_____

_____

8. What is spring back? Record your findings:

_____

_____

_____

INSTRUCTOR COMMENTS:

_____

_____

_____

_____

_____

_____

_____

Name _____   Date _____

Class _____   Instructor _____   Grade _____

1. *Technician A* says that a unibody vehicle should be anchored at the end of the front and rear rails to ensure that the entire vehicle is uniformly affected by the corrective forces. *Technician B* says that the area where the pinch weld clamps are attached should be finished, primed, and painted after removing them. Who is correct?
   A. Technician A only
   B. Technician B only
   C. Both Technicians A and B
   D. Neither Technician A nor B

2. *Technician A* says that pressure as a corrective force is more easily controlled than any other corrective force. *Technician B* says that the use of tension allows one to pull in a more precise direction than pressure. Who is correct?
   A. Technician A only
   B. Technician B only
   C. Both Technicians A and B
   D. Neither Technician A nor B

3. All of the following may be used as an attaching point when making corrective pulls EXCEPT:
   A. a flange
   B. a damaged suspension part
   C. an undamaged impact absorber
   D. damaged sheet metal

4. *Technician A* says that whenever tension is used as a corrective force, the entire area between the attaching point and the anchor point is affected equally. *Technician B* says that when tension is used, the pulling device will always attempt to seek its own center of the pulling line of force. Who is correct?
   A. Technician A only
   B. Technician B only
   C. Both Technicians A and B
   D. Neither Technician A nor B

5. During a collision, the inertial forces generally:
   A. cause damage away from the point of impact
   B. affect only areas immediately adjacent to the point of impact
   C. affect only nonstructural parts
   D. create damage that is readily visible and identifiable

6. *Technician A* says that the tie down/anchoring points on a unibody vehicle are the same as those of a BOF vehicle. *Technician B* says that the anchoring points are usually found under the four corners of the passenger compartment. Who is correct?
   A. Technician A only
   B. Technician B only
   C. Both Technicians A and B
   D. Neither Technician A nor B

7. Which of the following is true of using pressure as a corrective force?
   A. It is rarely used in conjunction with tension.
   B. It is readily directed to correct damage.
   C. It follows the path of least resistance.
   D. The direction it follows can easily be controlled.

8. *Technician A* says that a unibody vehicle must be tied down to a bench to form a frame because the vehicle has none. *Technician B* says that when making corrective pulls on a unibody, one must pull in only one direction at a time to avoid overpulling it and thus causing additional damage. Who is correct?
   A. Technician A only
   B. Technician B only
   C. Both Technicians A and B
   D. Neither Technician A nor B

9. The anchoring points used on the unibody vehicle must:
   A. be parallel to the direction of any pulls made when repairing the vehicle.
   B. be able to withstand the total of all the combined corrective forces used.
   C. be at the outermost part of the vehicle, front and back.
   D. be located at the front and rear of the passenger compartment.

10. *Technician A* says that stress relieving on high-strength steel surfaces must be done over a larger area than on standard or mild steel. *Technician B* says that a limited amount of heat can be used as long as the safe temperature time thresholds are not exceeded. Who is correct?
    A. Technician A only
    B. Technician B only
    C. Both Technicians A and B
    D. Neither Technician A nor B

11. *Technician A* says that an incorrect camber setting will cause a vehicle to wander in the direction of the least positive setting. *Technician B* says that caster is a stability angle. Who is correct?
    A. Technician A only
    B. Technician B only
    C. Both Technicians A and B
    D. Neither Technician A nor B

12. All of the following can be used as an attachment location for performing corrective pulls EXCEPT:
    A. damaged sheet metal that will be replaced
    B. the steering gear bolted to the frame rail
    C. a damaged inner fender rail that will be salvaged
    D. a collapsed B pillar

13. Which of the following is NOT considered a diagnostic angle?
    A. SAI
    B. included angle
    C. toe
    D. camber

14. *Technician A* says that a certain degree of thrust angle is desirable to keep the vehicle running straight. *Technician B* says that "0" running toe is the desired setting to achieve. Who is correct?
    A. Technician A only
    B. Technician B only
    C. Both Technicians A and B
    D. Neither Technician A nor B

15. *Technician A* says that when the damaged area is being pulled to restore it to the correct dimension, it may need to be pulled beyond its normal position to overcome the elastic factor. *Technician B* says that the point of impact will likely incur a considerable amount of inertial damage. Who is correct?
    A. Technician A only
    B. Technician B only
    C. Both Technicians A and B
    D. Neither Technician A nor B

Chapter 15

Structural Parts Replacement

SAFETY, ROUGH REPAIR, AND REMOVING PARTS

Name _____ Date _____

Class _____ Instructor _____ Grade _____

NATEF TASKS I A. 8, 9, 13; B. 17, 18, 19; II B 13; E. 1, 2, 4, 5, 6, 7, 8, 9, 10, 11, 12, 13,14, 15, 16, 17, 18, 19

1. After reading the assignment, in the space provided below, list the personal and environmental safety precautions necessary for this assignment.

   _____

   _____

   _____

2. Why are personal safety and vehicle safety important when working with structural parts? Record your findings:

   _____

   _____

   _____

3. If glass is not protected, what could happen? Record your findings:

   _____

   _____

   _____

4. List other parts of a vehicle that should be protected during repairs and explain why. Record your findings:

   _____

   _____

   _____

5. What is the importance of a repair plan? Record your findings:

   _____

   _____

   _____

6. What importance does inventorying parts serve? Why? Record your findings:

   _____

   _____

   _____

7. What is "bagging and tagging" and why is it important? Record your findings:

_____

_____

_____

8. What is rough pulling? Record your findings:

_____

_____

_____

9. How would a technician locate and remove factory spot welds? Record your findings:

_____

_____

_____

10. Explain how a spot weld is separated. Record your findings:

_____

_____

_____

11. List some parts that may be welded on or weld bonded on that are not structural. Record your findings:

_____

_____

_____

INSTRUCTOR COMMENTS:

_____

_____

_____

_____

_____

_____

UTILIZING RECOMMENDED REPAIR PROCEDURES AND WELDING SEQUENCES AND TECHNIQUES

Name _____ Date _____

Class _____ Instructor _____ Grade _____

NATEF TASKS I A. 8, 9, 13; B. 17, 18, 19; II B. 1; E. 1, 2, 4, 5, 6, 7, 8, 9, 10, 11, 12, 13,14, 15, 16, 17, 18, 19

1. After reading the assignment, in the space provided below, list the personal and environmental safety precautions necessary for this assignment.

   _____
   _____
   _____

2. Which method of quarter panel replacement is best, full or abbreviated? Explain why. Record your findings:

   _____
   _____
   _____

3. What is UPCR, and what can be found there? Record your findings:

   _____
   _____
   _____

4. Why is replacing a roof panel difficult on some vehicles? Record your findings:

   _____
   _____
   _____

5. What is a door skin, and how is it removed? Record your findings:

   _____
   _____
   _____

6. How is adhesive bonding used when installing a door skin? Is welding necessary? Record your findings:

   _____
   _____
   _____

7. What is meant by "factory seam replacement"? Record your findings:

   _____
   _____
   _____

8. What is a sectional replacement? Record your findings:

   _____
   _____
   _____

9. Describe the following joints:

Butt joint with a backer:

_____

_____

_____

Lap joint:

_____

_____

_____

Open butt:

_____

_____

_____

10. Deciding which joint to use and where it should be located is complex. How would a technician know the correct type and location? Record your findings:

_____

_____

_____

INSTRUCTOR COMMENTS:

_____

_____

_____

_____

_____

_____

UTILIZING RECOMMENDED ADHESIVE BONDING TECHNIQUES RESTORING FOAMS AND CORROSION PROTECTION

Name _____ Date _____

Class _____ Instructor _____ Grade _____

NATEF TASKS I A. 8, 9, 13; B. 17, 18, 19; II B. 13; E. 1, 2, 4, 5, 6, 7, 8, 9, 10, 11, 12, 13,14, 15, 16, 17, 18, 19

1. After reading the assignment, in the space provided below, list the personal and environmental safety precautions necessary for this assignment.

   _____

   _____

   _____

2. Why is restoring corrosion protection important? Record your findings:

   _____

   _____

   _____

3. Why is weld-through primer removed from the areas away from the joint following welding? Record your findings:

   _____

   _____

   _____

4. Where could adhesive bond be used in collision repair? Record your findings:

   _____

   _____

   _____

5. Compare and contrast adhesive bonding and weld bonding. Record your findings:

   _____

   _____

   _____

6. How would a technician know when to use bonding when repairing a vehicle? Record your findings:

   _____

   _____

   _____

7. How is the separation of a factory weld-bonded seam different from a spot-welded seam? Record your findings:

   _____

   _____

   _____

8. How does adhesive bonding affect corrosion protection? Record your findings:

   _____

   _____

INSTRUCTOR COMMENTS:

_____

_____

_____

_____

_____

_____

SAFETY, ROUGH REPAIR, AND REMOVING PARTS

Name _____   Date _____

Class _____   Instructor _____   Grade _____

OBJECTIVES

- Know and use safety techniques to protect both the vehicle and the technicians.
- Remove damaged cosmetic exterior parts, identifying and cataloging them and other peripheral parts and pieces.
- Rough repair damaged cosmetic and structural parts prior to removal and replacement with new panels.
- Locate and utilize recommended vehicle-specific repair procedures from a variety of sources.
- Identify the welding sequences and techniques used for permanent panel installation.
- Understand adhesive bonding versus welding and a combination of each method.
- Reinstall foam and corrosion protection.

NATEF TASK CORRELATION

The written and hands-on activities in this chapter satisfy the NATEF High Priority-Individual and High Priority-Group requirements for Sections I A. 8, 9, 13; B. 17, 18, 19; II B. 13; E. 1, 2, 4, 5, 6, 7, 8, 9, 10, 11, 12, 13,14, 15, 16, 17, 18, 19.

Tools and equipment needed (NATEF tool list)

- Safety glasses
- Ear protection
- Pencil and paper
- Damaged vehicle
- Vehicle service manual
- Welders
- Automotive foam
- Replacement parts
- Gloves
- Particle mask
- Assorted hand and body tools
- Estimate or work order for the vehicle
- Structural measuring equipment
- Assorted adhesive bonds
- Spot weld remover tools

Vehicle Description

Year_____ Make _____ Model _____

VIN _____ Paint Code _____

PROCEDURE

1. After reading the assignment, in the space provided below, list the personal and environmental safety equipment and precautions needed for this assignment. Have the instructor approve your list.

_____

_____

_____

INSTRUCTOR'S APPROVAL _____

Using the vehicle and work order/estimate provided by your instructor, complete the tasks below.

2. Check the vehicle, making sure that it is safe and that all necessary safety precautions have been performed. Record your findings:

   _____

   _____

   _____

3. After reading the repair order, make a repair plan for this vehicle. Record your plan:

   _____

   _____

   _____

4. Inventory and list the parts needed for the repair. Record your findings:

   _____

   _____

   _____

5. Remove for access and store all parts prior to repair. Bag, tag, and store the fasteners. Record your process:

   _____

   _____

   _____

6. Devise a rough pull and prepare it for the instructor's inspection prior to performing. Record your process:

   _____

   _____

   _____

7. Identify the factory spot welds that will need to be removed. How did you locate them? Record your findings:

   _____

   _____

   _____

8. Remove the spot welds. Record your procedure:

   _____

   _____

   _____

INSTRUCTOR COMMENTS:

_____

_____

_____

_____

_____

_____

UTILIZING RECOMMENDED REPAIR PROCEDURES AND WELDING SEQUENCES AND TECHNIQUES

Name _____   Date _____

Class _____   Instructor _____   Grade _____

OBJECTIVES

- Know and use safety techniques to protect both the vehicle and the technicians.
- Remove damaged cosmetic exterior parts, identifying and cataloging them and other peripheral parts and pieces.
- Rough repair damaged cosmetic and structural parts prior to removal and replacement with new panels.
- Locate and utilize recommended vehicle-specific repair procedures from a variety of sources.
- Identify the welding sequences and techniques used for permanent panel installation.
- Understand adhesive bonding versus welding and a combination of each method.
- Reinstall foam and corrosion protection.

NATEF TASK CORRELATION

The written and hands-on activities in this chapter satisfy the NATEF High Priority-Individual and High Priority-Group requirements for Sections I A. 8, 9, 13; B. 17, 18, 19; II B. 13; E. 1, 2, 4, 5, 6, 7, 8, 9, 10, 11, 12, 13,14, 15, 16, 17, 18, 19.

Tools and equipment needed (NATEF tool list)

- Safety glasses
- Ear protection
- Pencil and paper
- Damaged vehicle
- Vehicle service manual
- Welders
- Automotive foam
- Replacement parts
- Gloves
- Particle mask
- Assorted hand and body tools
- Estimate or work order for the vehicle
- Structural measuring equipment
- Assorted adhesive bonds
- Spot weld remover tools

Vehicle Description

Year_____ Make _____ Model _____

VIN _____ Paint Code _____

PROCEDURE

1. After reading the assignment, in the space provided below, list the personal and environmental safety equipment and precautions needed for this assignment. Have the instructor approve your list.

_____

_____

_____

INSTRUCTOR'S APPROVAL _____

Using the vehicle and work order/estimate provided by your instructor, complete the tasks below.

2. Using the UPCR, find the recommended repair procedure for this repair Record your findings:

_____

_____

_____

3. List your repair plan:

_____

_____

_____

4. Will the repair require either adhesive bonding or weld-through bonding? If so, explain which bonding agent you will use and why. Record your findings:

_____

_____

_____

5. Will the repair require foam? If so, which agent should you use? Record your findings:

_____

_____

_____

6. Prepare the mating surfaces for fit-up and welding, including corrosion protection. Record your procedure:

_____

_____

_____

7. Prepare the weld joint or mating surfaces and have the instructor inspect it before fit-up. Record your findings:

_____

_____

_____

8. Fit up the replacement part and have the instructor check it prior to welding or bonding. Record your findings:

_____

_____

_____

INSTRUCTOR COMMENTS:

_____

_____

_____

_____

_____

UTILIZING RECOMMENDED ADHESIVE BONDING TECHNIQUES RESTORING FOAMS AND CORROSION PROTECTION

Name _____ Date _____

Class _____ Instructor _____ Grade _____

OBJECTIVES

- Know and use safety techniques to protect both the vehicle and the technicians.
- Remove damaged cosmetic exterior parts, identifying and cataloging them and other peripheral parts and pieces.
- Rough repair damaged cosmetic and structural parts prior to removal and replacement with new panels.
- Locate and utilize recommended vehicle-specific repair procedures from a variety of sources.
- Identify the welding sequences and techniques used for permanent panel installation.
- Understand adhesive bonding versus welding and a combination of each method.
- Reinstall foam and corrosion protection.

NATEF TASK CORRELATION

The written and hands-on activities in this chapter satisfy the NATEF High Priority-Individual and High Priority-Group requirements for Sections I A. 8, 9, 13; B. 17, 18, 19; II B. 13; E. 1, 2, 4, 5, 6, 7, 8, 9, 10, 11, 12, 13,14, 15, 16, 17, 18, 19.

Tools and equipment needed (NATEF tool list)

- Safety glasses
- Ear protection
- Pencil and paper
- Damaged vehicle
- Vehicle service manual
- Welders
- Automotive foam
- Replacement parts
- Gloves
- Particle mask
- Assorted hand and body tools
- Estimate or work order for the vehicle
- Structural measuring equipment
- Assorted adhesive bonds
- Spot weld remover tools

Vehicle Description

Year_____ Make _____ Model _____

VIN _____ Paint Code _____

PROCEDURE

1. After reading the assignment, in the space provided below, list the personal and environmental safety equipment and precautions needed for this assignment. Have the instructor approve your list.

_____

_____

_____

INSTRUCTOR'S APPROVAL _____

Using the vehicle and work order/estimate provided by your instructor, complete the tasks below.

2. Prepare a practice butt joint with a backer, and have it approved by the instructor. Record your findings:

_____

_____

_____

INSTRUCTOR'S APPROVAL _____

3. Prepare a practice butt joint and have it approved by the instructor. Record your findings:

_____

_____

_____

INSTRUCTOR'S APPROVAL _____

4. Prepare a practice lap joint with a backer and have it approved by the instructor. Record your findings:

_____

_____

_____

INSTRUCTOR'S APPROVAL _____

5. Prepare a practice weld bond joint and have it approved by the instructor. Record your findings:

_____

_____

_____

INSTRUCTOR'S APPROVAL _____

6. Using scrap steel from the vehicle being repaired, make practice welds to fine-tune the welder. Have your instructor check the finished welds. Record your findings:

_____

_____

_____

INSTRUCTOR'S APPROVAL _____

7. Weld the fit-up part and have your instructor approve the finished welds. Record your findings:

_____

_____

_____

INSTRUCTOR'S APPROVAL _____

INSTRUCTOR COMMENTS:

_____

_____

_____

_____

_____

1. *Technician A* says that a straight cut open butt joint can be used to resection a C pillar when making a sectional replacement. *Technician B* says that an insert will help to strengthen the area where the joint is welded. Who is correct?
   A. Technician A only
   B. Technician B only
   C. Both Technicians A and B
   D. Neither Technician A nor B

2. *Technician A* says that a tapered lap joint may be used in certain areas if there is not enough of a smooth and uninterrupted surface to insert a sleeve. *Technician B* says that welding the trunk floor to the open side of a "U" channel creates a hat channel. Who is correct?
   A. Technician A only
   B. Technician B only
   C. Both Technicians A and B
   D. Neither Technician A nor B

3. *Technician A* says that using recycled parts can be advantageous because the recycler may include extra parts that may require replacing but would not generally be included with new parts. *Technician B* says that a uniside panel may include the quarter panel, the center pillar, and door openings. Who is correct?
   A. Technician A only
   B. Technician B only
   C. Both Technicians A and B
   D. Neither Technician A nor B

4. *Technician A* says that the flanged or recessed lap joint is commonly used when performing a sectional rail replacement. *Technician B* says that the abbreviated quarter panel replacement method usually includes replacing the sail panel. Who is correct?
   A. Technician A only
   B. Technician B only
   C. Both Technicians A and B
   D. Neither Technician A nor B

5. *Technician A* says that when installing a replacement center or B pillar, the technician should add an additional 30% more MIG plug welds than the manufacturer originally used. *Technician B* says that increasing the number of MIG plug welds that the manufacturer used may alter the vehicle's ability to distribute the impact forces in a subsequent collision. Who is correct?
   A. Technician A only
   B. Technician B only
   C. Both Technicians A and B
   D. Neither Technician A nor B

6. *Technician A* says that before any part of the damaged structural part is removed, it must be roughed out to relieve the strain from the area. *Technician B* says that all door and window openings must be roughly restored before removing any part of the structural member. Who is correct?
   A. Technician A only
   B. Technician B only
   C. Both Technicians A and B
   D. Neither Technician A nor B

7. *Technician A* says that when welding the replacement structure into place, the technician should make the first welds on the definition crowns. *Technician B* says that a stitch weld technique should be used when welding the seam to minimize the possibility of warping. Who is correct?
   A. Technician A only
   B. Technician B only
   C. Both Technicians A and B
   D. Neither Technician A nor B

8. *Technician A* says that a sectional replacement can be made by cutting through the center of the strut tower and overlapping the replacement part over the old tower. *Technician B* says that it is advantageous to use an insert for reinforcement any time it is possible. Who is correct?
   A. Technician A only
   B. Technician B only
   C. Both Technicians A and B
   D. Neither Technician A nor B

9. *Technician A* says that when drilling out spot welds, one should drill through both surfaces, leaving a hole where the weld once was located. *Technician B* says that using the air hammer will separate the metal surfaces more quickly and cleanly than drilling the spot welds first. Who is correct?
   A. Technician A only
   B. Technician B only
   C. Both Technicians A and B
   D. Neither Technician A nor B

10. *Technician A* says that when performing a full factory replacement of the B pillar, the outer skin of the roof must also be replaced. *Technician B* says that the best location for the seam is between the D ring anchor point and the belt line when performing a sectional replacement on the B pillar. Who is correct?
    A. Technician A only
    B. Technician B only
    C. Both Technicians A and B
    D. Neither Technician A nor B

11. *Technician A* says that when making a sectional replacement of the A pillar, a staggered butt may be used. *Technician B* says that a butt joint with insert may be used. Who is correct?
    A. Technician A only
    B. Technician B only
    C. Both Technicians A and B
    D. Neither Technician A nor B

12. *Technician A* says that weld bonding can be used on all structural panels, provided adequate curing time is allowed. *Technician B* says that weld bonding affords better corrosion protection in the seam area than other approved methods. Who is correct?
    A. Technician A only
    B. Technician B only
    C. Both Technicians A and B
    D. Neither Technician A nor B

13. *Technician A* says that any time the door intrusion beam is bent, the door assembly should be replaced. *Technician B* says that the door outer repair panel can be installed using either adhesives and crimping the hemming flange or by crimping the edge and using small spot welds around the outer edge to hold the skin in place. Who is correct?
    A. Technician A only
    B. Technician B only
    C. Both Technicians A and B
    D. Neither Technician A nor B

14. *Technician A* says that silicon bronze is used in some seams to increase the flexibility of the joint without compromising strength. *Technician B* says that when using STRSW, one should increase the number of welds that the manufacturer used by approximately 30%. Who is correct?
    A. Technician A only
    B. Technician B only
    C. Both Technicians A and B
    D. Neither Technician A nor B

15. *Technician A* says that a full body section involves cutting through the A, B, and C pillars and across the floor and rocker panels. *Technician B* says that an open butt weld joint should be made on both the B pillar and the C pillar when welding them together during a full body section. Who is correct?
    A. Technician A only
    B. Technician B only
    C. Both Technicians A and B
    D. Neither Technician A nor B

Chapter 16

Full Frame Sectioning and Replacement

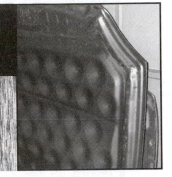

■ WORK ASSIGNMENT 16-1

FRAME DESIGN

Name _____ Date _____

Class _____ Instructor _____ Grade _____

NATEF TASKS A 1, 2, 3, 4, 5, 6, 7, 9, 10, 11, 12, 16, 17; B 1, 6, 8, 15, 16, 17, 18, 19, 21

1. After reading the assignment, in the space provided below, list the personal and environmental safety equipment and precautions needed for this assignment.

   _____

   _____

   _____

2. Describe the importance of recoil straps and how they are used when repairing BOF damage. Record your findings:

   _____

   _____

   _____

3. List any additional safety precautions that should be observed when using frame repair equipment and explain why. Record your findings:

   _____

   _____

   _____

4. Briefly describe the frame types listed below and explain their usage.

   Ladder frame:

   _____

   _____

   _____

   Perimeter frame:

   _____

   _____

   _____

   Stub frame:

   _____

   _____

   _____

5. What repair considerations should be observed when repairing a BOF vs. a unibody vehicle? Record your findings:

_____

_____

_____

6. What, role, if any, does heat have in the repair of BOF vehicles? Record your findings:

_____

_____

_____

INSTRUCTOR COMMENTS:

_____

_____

_____

_____

_____

_____

■ WORK ASSIGNMENT 16-2

HEATING, SECTIONING AND REPLACING BOF

Name _____ Date _____

Class _____ Instructor _____ Grade _____

$

NATEF TASK A 1, 2, 3, 4, 5, 6, 7, 9, 10, 11, 12, 16, 17; B 1, 6, 8, 15, 16, 17, 18, 19, 21

1. After reading the assignment, in the space provided below, list the personal and environmental safety equipment and precautions needed for this assignment.

   _____

   _____

   _____

2. What roles do heat, the amount of heat, and its duration have when repairing framed vehicles? Record your findings:

   _____

   _____

   _____

3. List and describe the methods used to monitor heat. Record your findings:

   _____

   _____

   _____

4. Where or how would a technician determine the correct temperature or recommended duration of heat for a specific frame? Record your findings:

   _____

   _____

   _____

5. Describe the Kink vs. Bent theory of replacement. Record your findings:

   _____

   _____

   _____

6. What is a frame replacement, and when would it be necessary? Record your findings:

   _____

   _____

   _____

7. When repairing a BOF, how would it be prepared, anchored, and rough pulled? Record your findings:

   _____

   _____

8. How and when would a frame section be replaced on a BOF vehicle? Record your findings:

   _____

   _____

   _____

9. What is an OEM partial frame section, and what is it used for? Record your findings:

_____

_____

_____

10. What are general sectioning guidelines, and how could one find them? Record your findings:

_____

_____

_____

INSTRUCTOR COMMENTS:

_____

_____

_____

_____

_____

_____

_____

REPAIR TECHNIQUES

Name _____ Date _____

Class _____ Instructor _____ Grade _____

$

NATEF TASKS A 1, 2, 3, 4, 5, 6, 7, 9, 10, 11, 12, 16, 17; B 1, 6, 8, 15, 16, 17, 18, 19, 21

1. After reading the assignment, in the space provided below, list the personal and environmental safety equipment and precautions needed for this assignment.

   _____

   _____

   _____

2. In the spaces below, briefly describe and explain the use of each of these items.

   Inserts:

   _____

   _____

   _____

   Offset butt:

   _____

   _____

   _____

   Offset fillet or lap joint:

   _____

   _____

   _____

3. What are stress and fatigue cracks, and how are they repaired?

   _____

   _____

   _____

4. Why should a vehicle be protected from further damage during the rep air process?

   _____

   _____

   _____

5. How should it be protected?

   _____

   _____

   _____

6. Why are structurally sound welds so important with BOF?

   _____

   _____

   _____

7. How is the introduction of aluminum in BOF changing repair techniques and shops?

_____

_____

_____

8. How does developing a work plan increase repair quality and productivity?

_____

_____

_____

9. What special repair guidelines must a technician be aware of when repairing a BOF vehicle?

_____

_____

_____

_____

INSTRUCTOR COMMENTS:

_____

_____

_____

_____

_____

_____

FRAME DESIGN

Name _____   Date _____

Class _____   Instructor _____   Grade _____

OBJECTIVES

- Identify the various frame designs.
- Discuss the most common applications for each frame design.
- Identify the types of metals used and the repair and replacement variables for mild, HSS, HSLA, and UHSS steel.
- Follow the recommendations for proper and correct heat applications.
- Understand considerations for sectional replacement.
- Understand considerations for full frame replacement.
- Develop a repair plan for replacing the damaged section or the entire frame assembly.
- Identify the variables and considerations for repairing and replacing frames, utilizing recently introduced technology.
- Distinguish different repair techniques used for repairing and replacing aluminum structural and frame sections.
- Follow safety practices to protect both the vehicle and the technicians.

NATEF TASK CORRELATION

The written and hands-on activities in this chapter satisfy the NATEF High Priority-Individual and High Priority-Group requirements for Tasks A 1, 2, 3, 4, 5, 6, 7, 9, 10, 11, 12, 16, 17; B 1, 6, 8, 15, 16, 17 ,18, 19, 21.

Tools and equipment needed (NATEF tool list)

- Safety glasses
- Ear protection
- Pencil and paper
- Damaged vehicle
- Vehicle service manual
- Pulling equipment
- Gloves
- Particle mask
- Assorted hand tools
- Estimate or work order for the vehicle
- Structural measuring equipment

Vehicle Description

Year_____ Make _____ Model _____

VIN _____ Paint Code _____

PROCEDURE

1. After reading the assignment, in the space provided below, list the personal and environmental safety equipment and precautions needed for this assignment. Have the instructor review and approve your list.

_____

_____

_____

INSTRUCTOR'S APPROVAL _____

Using the vehicle and work order/estimate provided by your instructor, complete the tasks below.

2. Identify and describe the damage on a ladder frame vehicle. Record your findings:

_____

_____

3. Identify and describe the damage on a subframe vehicle. Record your findings:

_____

_____

4. Identify and describe the damage on a perimeter frame vehicle. Record your findings:

_____

_____

5. On the vehicle provided by your instructor, identify the types of steel used to manufacture the frame, and determine its heat parameters. Record your findings:

_____

_____

_____

INSTRUCTOR COMMENTS:

_____

_____

_____

_____

_____

_____

HEATING, SECTIONING, AND REPLACING BOF

Name _____ Date _____

Class _____ Instructor _____ Grade _____

OBJECTIVES

- Identify the various frame designs.
- Discuss the most common applications for each frame design.
- Identify the types of metals used and the repair and replacement variables for mild, HSS, HSLA, and UHSS steel.
- Follow the recommendations for proper and correct heat applications.
- Understand considerations for sectional replacement.
- Understand considerations for full frame replacement.
- Develop a repair plan for replacing the damaged section or the entire frame assembly.
- Identify the variables and considerations for repairing and replacing frames, utilizing recently introduced technology.
- Distinguish different repair techniques used for repairing and replacing aluminum structural and frame sections.
- Follow safety practices to protect both the vehicle and the technicians.

NATEF TASK CORRELATION

The written and hands-on activities in this chapter satisfy the NATEF High Priority-Individual and High Priority-Group requirements for Tasks A 1, 2, 3, 4, 5, 6, 7, 9, 10, 11, 12, 16, 17; B 1, 6, 8, 15, 16, 17, 18, 19, 21.

Tools and equipment needed (NATEF tool list)

- Safety glasses
- Ear protection
- Pencil and paper
- Damaged vehicle
- Vehicle service manual
- Pulling equipment
- Gloves
- Particle mask
- Assorted hand tools
- Estimate or work order for the vehicle
- Structural measuring equipment

Vehicle Description

Year_____ Make _____ Model _____

VIN _____ Paint Code _____

PROCEDURE

1. After reading the assignment, in the space provided below, list the personal and environmental safety equipment and precautions needed for this assignment. Have the instructor review and approve your list.

_____

_____

_____

INSTRUCTOR'S APPROVAL _____

Using the vehicle and work order/estimate provided by your instructor, complete the tasks below.

2. On the vehicle provided, determine the heat restrictions that should be followed. Record your findings:

_____

_____

_____

3. After assesing the vehicle provided, determine if the parts can be repaired or should be replaced. Record your findings:

_____

_____

_____

4. On the vehicle provided, mount anchor and prepare for a rough pull. Have the instructor approve the setup before proceeding:

INSTRUCTOR'S APPROVAL _____

5. On the prepared and approved vehicle, make a rough pull using the heat guidelines listed earlier. Record your process:

_____

_____

_____

INSTRUCTOR COMMENTS:

_____

_____

_____

_____

_____

_____

REPAIR TECHNIQUES

Name _____   Date _____

Class _____   Instructor _____   Grade _____

OBJECTIVES

- Identify the various frame designs.
- Discuss the most common applications for each frame design.
- Identify the types of metals used and the repair and replacement variables for mild, HSS, HSLA, and UHSS steel.
- Follow the recommendations for proper and correct heat applications.
- Understand considerations for sectional replacement.
- Understand considerations for full frame replacement.
- Develop a repair plan for replacing the damaged section or the entire frame assembly.
- Identify the variables and considerations for repairing and replacing frames, utilizing recently introduced technology.
- Distinguish different repair techniques used for repairing and replacing aluminum structural and frame sections.
- Follow safety practices to protect both the vehicle and the technicians.

NATEF TASK CORRELATION

The written and hands-on activities in this chapter satisfy the NATEF High Priority-Individual and High Priority-Group requirements for Tasks A 1, 2, 3, 4, 5, 6, 7, 9, 10, 11, 12, 16, 17; B 1, 6, 8, 15, 16, 17, 18, 19, 21.

Tools and equipment needed (NATEF tool list)

- Safety glasses
- Ear protection
- Pencil and paper
- Damaged vehicle
- Vehicle service manual
- Pulling equipment
- Gloves
- Particle mask
- Assorted hand tools
- Estimate or work order for the vehicle
- Structural measuring equipment

Vehicle Description

Year_____ Make _____ Model _____

VIN _____ Paint Code _____

PROCEDURE

1. After reading the assignment, in the space provided below, list the personal and environmental safety equipment and precautions needed for this assignment. Have the instructor review and approve your list.

   _____

   _____

   _____

INSTRUCTOR'S APPROVAL _____

Using the vehicle and work order/estimate provided by your instructor, complete the tasks below.

2. For the vehicle you previously roughly pulled, make a repair plan for sectioning the frame. Record your plan and have the instructor approve:

_____

_____

_____

INSTRUCTOR'S APPROVAL _____

3. Prepare the new part and the frame for welding.

INSTRUCTOR'S APPROVAL _____

4. Weld the new part to the frame.

INSTRUCTOR'S APPROVAL _____

INSTRUCTOR COMMENTS:

_____

_____

_____

_____

_____

_____

Name _____   Date _____

Class _____   Instructor _____   Grade _____

1. *Technician A* says that all frame rails are made of common low carbon mild steel. *Technician B* says that the frame rails are made from several steel grades. Who is correct?
   A. Technician A only
   B. Technician B only
   C. Both Technicians A and B
   D. Neither Technician A nor B

2. *Technician A* says that heat can safely be applied to high-strength steel as long as the upper temperature threshold of 1,500°F is not exceeded. *Technician B* says that heat can be applied for 3 minutes cumulatively as long as it is not heated to more than 1,200°F. Who is correct?
   A. Technician A only
   B. Technician B only
   C. Both Technicians A and B
   D. Neither Technician A nor B

3. *Technician A* says that the ladder frame is usually used on large heavy-duty trucks. *Technician B* says that the ladder frame is used because it is stronger than any other type of frame design. Who is correct?
   A. Technician A only
   B. Technician B only
   C. Both Technicians A and B
   D. Neither Technician A nor B

4. *Technician A* says that before a damaged frame section can be removed from the rest of the frame, it must be roughly repaired to relieve the stresses. *Technician B* says that a butt joint with insert is commonly used when an enclosed box rail section is replaced. Who is correct?
   A. Technician A only
   B. Technician B only
   C. Both Technicians A and B
   D. Neither Technician A nor B

5. *Technician A* says that the manufacturer uses crush zones to help manage the collision energy. *Technician B* says that if vehicle-specific repair procedures are not available, one should use the general sectioning guidelines provided by Tech Cor and I-CAR. Who is correct?
   A. Technician A only
   B. Technician B only
   C. Both Technicians A and B
   D. Neither Technician A nor B

6. *Technician A* says that the root gap should be approximately the equivalent of two thicknesses of the metal in the frame rail when making a butt joint with an insert. *Technician B* says that two welding passes may have to be made when welding the new rail onto the existing one. Who is correct?
   A. Technician A only
   B. Technician B only
   C. Both Technicians A and B
   D. Neither Technician A nor B

7. *Technician A* says that the SMAW, MIG, or flux core welding process must be used for welding on nearly all frame rails. *Technician B* says that ideally the electronic control module should be removed from the vehicle prior to making any MIG welds. Who is correct?
   A. Technician A only
   B. Technician B only
   C. Both Technicians A and B
   D. Neither Technician A nor B

8. *Technician A* says that the backside of the rail must always be reinforced with another layer of metal whenever a butt joint is used to secure a rail section. *Technician B* says that a 110V to 115V welder is adequate for performing all the required welds when replacing frame sections. Who is correct?
   A. Technician A only
   B. Technician B only
   C. Both Technicians A and B
   D. Neither Technician A nor B

9. *Technician A* says that an offset fillet weld is usually used when one section of the rail is slid inside another one. *Technician B* says that high-strength steel and mild steel may be used for parts that are hydroformed. Who is correct?
   A. Technician A only
   B. Technician B only
   C. Both Technicians A and B
   D. Neither Technician A nor B

10. *Technician A* says that some manufacturers use sacrificial parts at the outer end of the rail to protect the rail deeper into the vehicle and to help absorb some of the collision energy. *Technician B* says that disconnecting the battery will help to protect the electrical system. Who is correct?
    A. Technician A only
    B. Technician B only
    C. Both Technicians A and B
    D. Neither Technician A nor B

11. *Technician A* says that an advantage of making a sectional replacement is that it requires less disassembling of the vehicle. *Technician B* says that most OEM frame assemblies sold as replacement parts for damaged vehicles are usually made for the standard model. Who is correct?
    A. Technician A only
    B. Technician B only
    C. Both Technicians A and B
    D. Neither Technician A nor B

12. *Technician A* says that a hydroformed part is usually recognizable because it has sharply formed square corners. *Technician B* says that a kinked frame rail should be replaced because it cannot be repaired without the use of heat. Who is correct?
    A. Technician A only
    B. Technician B only
    C. Both Technicians A and B
    D. Neither Technician A nor B

13. *Technician A* says that a noncontact thermometer is the only means used to monitor the temperature of the metal when applying heat. *Technician B* says that plug welds are used to hold the insert into place when the two rails are aligned. Who is correct?
    A. Technician A only
    B. Technician B only
    C. Both Technicians A and B
    D. Neither Technician A nor B

14. *Technician A* says that the perimeter frame uses a drop center design to assist in managing the collision energy. *Technician B* says that placing seams at suspension mounting locations should be avoided when making a sectional replacement. Who is correct?
    A. Technician A only
    B. Technician B only
    C. Both Technicians A and B
    D. Neither Technician A nor B

15. *Technician A* says that even though the General Sectioning Guidelines are not vehicle specific, they can be used for making most repairs safely. *Technician B* says that placing a seam in a crush zone will not affect the timing of an air bag deployment. Who is correct?
    A. Technician A only
    B. Technician B only
    C. Both Technicians A and B
    D. Neither Technician A nor B

Chapter 17

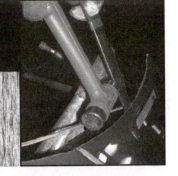

Welded Exterior Panel Replacement

■ WORK ASSIGNMENT 17-1

ANALYSIS OF STRUCTURAL DAMAGE AND PERSONAL SAFETY

Name _____ Date _____

Class _____ Instructor _____ Grade _____

1. After reading the assignment, in the space provided below, list the personal and environmental safety equipment and precautions needed for this assignment.

    _____

    _____

    _____

2. In your own words explain the kink vs. bent theory.

    _____

    _____

3. List below how you would identify a kinked vehicle part.

    _____

    _____

4. List below how you would identify a bent vehicle part.

    _____

    _____

    _____

5. Explain what a structural part is.

    _____

    _____

6. How would you identify a structural part on a vehicle?

    _____

    _____

    _____

7. Explain if or why a kinked structural part that will be taken off the vehicle and replaced with a new part may need to be pulled.

    _____

    _____

    _____

8. Explain the importance of precise, accurate, and multiple measurements when replacing structural parts.

_____

_____

_____

INSTRUCTOR COMMENTS:

_____

_____

_____

_____

_____

_____

_____

DAMAGE ANALYSES

Name _____   Date _____

Class _____   Instructor _____   Grade _____

NATEF STRUCTURAL ANALYSIS AND DAMAGE REPAIR SECTION I A. 11; B. 2, 7, 18

1. After reading the assignment, in the space provided below, list the personal and environmental safety equipment and precautions needed for this assignment.

_____

_____

_____

2. Using the vehicle provided, list which parts will be replaced and which ones will be repaired.

_____

_____

_____

3. List each part that you believe should be replaced and give your rationale.

_____

_____

_____

4. List the parts that should be replaced at factory seams and why.

   1 _____

   _____

   2 _____

   _____

   3 _____

   _____

   4 _____

   _____

5. List the parts that will not be replaced at factory seams and your rationale why.

   1 _____

   _____

   2 _____

   _____

   3 _____

   _____

   4 _____

   _____

6. What is sectioning and when is it the best repair procedure?

_____

_____

_____

INSTRUCTOR COMMENTS:

_____

_____

_____

_____

_____

_____

STRUCTURAL JOINTS

Name _____ Date _____

Class _____ Instructor _____ Grade _____

1. After reading the assignment, in the space provided below, list the personal and environmental safety equipment and precautions needed for this assignment.

   _____

   _____

   _____

2. Identify and describe a lap joint.

   _____

   _____

3. Identify and describe an offset butt joint.

   _____

   _____

4. Identify and describe an open butt joint.

   _____

   _____

5. Identify and describe a butt joint with a backer joint.

   _____

   _____

6. Describe how to prepare a lap joint.

   _____

   _____

7. Describe how to prepare an offset butt joint.

   _____

   _____

8. Describe how to prepare an open butt joint.

   _____

   _____

9. Describe how to prepare an open butt joint.

   _____

   _____

INSTRUCTOR COMMENTS:

_____

_____

_____

_____

_____

_____

WELDING AND BONDING EXTERIOR PARTS

Name _____ Date _____

Class _____ Instructor _____ Grade _____

1. After reading the assignment, in the space provided below, list the personal and environmental safety equipment and precautions needed for this assignment.

_____

_____

_____

2. Using the joints from Work Order 17-3, describe which types, their location, and number of welds will be used to join them. List your conclusions.

Lap joint:

_____

_____

Offset butt joint:

_____

_____

Open butt joint:

_____

_____

Butt with a backer joint:

_____

_____

3. Describe the steps for adhesive bonding a joint.

_____

_____

_____

4. List the steps for welding a panel in place.

_____

_____

_____

INSTRUCTOR COMMENTS:

_____

_____

_____

_____

_____

_____

ANALYSIS OF STRUCTURAL DAMAGE AND PERSONAL SAFETY
(NATEF SECTION 1 A. 1, 8, 12, B. 1, 3, 21)

Name _____ Date _____

Class _____ Instructor _____ Grade _____

OBJECTIVES

- Know, understand, and use the safety equipment necessary for the task.
- Analyze the damage and identify a kinked part and a bent part.
- Analyze the damage and identify the structural and nonstructural parts on a vehicle and know where to find the information that identifies structural parts for specific vehicles.
- Demonstrate why precise measurements are vital when analyzing structural damage.

NATEF TASK CORRELATION

The written and hands-on activities in this chapter satisfy the NATEF High Priority-Individual and High Priority-Group requirements for Structural Analysis and Damage Repair Section I A. 1, 8, 12; B. 1, 3, 21.

Tools and equipment needed (NATEF tool list)

- Safety glasses
- Ear protection
- Pencil and paper
- Welding helmet
- Vehicle measuring equipment
- Tram gauge

- Gloves
- Particle mask
- Assorted hand tools
- Welding protective clothing
- Measuring tape
- Self-centering gauges or electronic measuring system

Vehicle Description

Year_____ Make _____ Model _____

VIN _____ Paint Code _____

PROCEDURE

1. After reading the work order, gather the safety gear needed to complete the task. In the space provided below, list the personal and environmental safety equipment and precautions needed for this assignment. Have the instructor check and approve your plan before proceeding.

_____

_____

_____

INSTRUCTOR'S APPROVAL _____

2. On the vehicle provided, analyze the damage and make a repair plan to measure the vehicle. Record your findings:

_____

_____

_____

3. In the space below, list any parts that may need to be removed to gain access to accurately assess the damage.

_____

_____

_____

4. Will a rough pull be needed to gain access? In the space below list your plan.

_____

_____

_____

5. Identify and list the kinked parts.

_____

_____

_____

6. Identify the bent parts. Record your findings:

_____

_____

_____

7. With the measuring tools provided, measure the vehicle to assess full body damage. Record your findings:

_____

_____

_____

8. Make a rough pull if necessary and remeasure. Record your findings:

_____

_____

_____

9. Is it necessary to measure again? If so, explain.

_____

_____

_____

10. List the bent or kinked parts of this vehicle that are considered structural.

_____

_____

_____

11. Prepare a repair plan for this vehicle.

_____

_____

_____

INSTRUCTOR COMMENTS:

_____

_____

_____

_____

_____

_____

DAMAGE ANALYSES

Name _____   Date _____

Class _____   Instructor _____   Grade _____

OBJECTIVES

- Know, understand, and use the safety equipment necessary for the task.
- Identify what items must be taken into consideration when deciding to replace or repair a structural part.
- Demonstrate a part replaced at factory seams.
- Demonstrate a part that is sectioned.

NATEF TASK CORRELATION

The written and hands-on activities in this chapter satisfy the NATEF High Priority-Individual and High Priority-Group requirements for Structural Analysis and Damage Repair Section I A. 11; B. 2, 7, 18.

Tools and equipment needed (NATEF tool list)

- Safety glasses
- Ear protection
- Pencil and paper
- Welding helmet
- Vehicle measuring equipment
- Tram gauge

- Gloves
- Particle mask
- Assorted hand tools
- Welding protective clothing
- Measuring tape
- Self-centering gauges or electronic measuring system

Vehicle Description

Year_____ Make _____ Model _____

VIN _____ Paint Code _____

PROCEDURE

1. After reading the work order, gather the safety gear needed to complete the task. In the space provided below, list the personal and environmental safety equipment and precautions needed for this assignment. Have the instructor check and approve your plan before proceeding.

   _____

   _____

   _____

INSTRUCTOR'S APPROVAL _____

2. Using the vehicle provided, list the steps you will need to take to analyze the damage.

   _____

   _____

3. Will parts need to be removed before the vehicle can be accurately analyzed? If so list them.

   _____

   _____

4. Make an initial measurement. Record your findings:

_____

_____

_____

5. Does the vehicle need to be rough pulled? Record your findings:

_____

_____

_____

6. Identify the parts that will need to be replaced. Make a repair plan for each, including which joint will be used and the welding procedure needed. Record your findings:

1 _____

_____

2 _____

_____

3 _____

_____

4 _____

_____

7. Identify the parts that will need to be repaired. Make a repair plan for each.

1 _____

_____

2 _____

_____

3 _____

_____

4 _____

_____

INSTRUCTOR COMMENTS:

_____

_____

_____

_____

_____

_____

STRUCTURAL JOINTS

Name _____ Date _____

Class _____ Instructor _____ Grade _____

OBJECTIVES

- Know, understand ,and use the safety equipment necessary for the task.
- Determine which of the four main types of joints should be used.
- Prepare and fit up a lap joint, offset butt joint, open butt joint, and butt joint with a backer.

NATEF TASK CORRELATION

The written and hands-on activities in this chapter satisfy the NATEF High Priority-Individual and High Priority-Group requirements.

Tools and equipment needed (NATEF tool list)

- Safety glasses
- Ear protection
- Pencil and paper
- Cutting tools
- Welding safety equipment
- Template paper
- Vehicle description

- Gloves
- Particle mask
- Assorted hand tools
- Welder
- Measuring tools
- Marker

Vehicle Description

Year_____ Make _____ Model _____

VIN _____ Paint Code _____

PROCEDURE

1. After reading the work order, gather the safety gear needed to complete the task. In the space provided below, list the personal and environmental safety equipment and precautions needed for this assignment. Have the instructor check and approve your plan before proceeding.

   _____
   _____
   _____

INSTRUCTOR'S APPROVAL _____

2. On a practice rail measure, cut, make, and fit up a lap joint. List the procedure below.

   _____
   _____

3. On a practice rail measure, cut, make, and fit up an offset butt joint. List the procedure below.

   _____
   _____
   _____

4. On a practice rail measure, cut, make, and fit up an open butt joint. List the procedure below.

_____

_____

_____

5. On a practice rail measure, cut, make, and fit up a butt joint with a backer joint. List the procedure below.

_____

_____

_____

INSTRUCTOR COMMENTS:

_____

_____

_____

_____

_____

_____

WELDING AND BONDING EXTERIOR PARTS

Name _____   Date _____

Class _____   Instructor _____   Grade _____

OBJECTIVES

- Know, understand, and use the safety equipment necessary for the task.
- Demonstrate the welding of spot welds, welded joints, and bonded joints.
- Demonstrate the welding of a lap joint, offset joint, open butt joint, and a butt joint with a backer.
- Find specific or general vehicle specifications for replacing the following items: door skins, quarter panels, a vehicle roof, and rear body panels.

NATEF TASK CORRELATION

The written and hands-on activities in this chapter satisfy the NATEF High Priority-Individual and High Priority-Group requirements.

Tools and equipment needed (NATEF tool list)

- Safety glasses
- Ear protection
- Pencil and paper
- Cutting tools
- Welding safety equipment
- Template paper

- Gloves
- Particle mask
- Assorted hand tools
- Welder
- Measuring tools
- Marker

Vehicle Description

Year_____ Make _____ Model _____

VIN _____ Paint Code _____

PROCEDURE

1. In the space provided below, list the personal and environmental safety precautions necessary for this assignment.

   _____

   _____

   _____

2. Using welding coupons provided, make 10 expectable:
   Spot welds:

   Instructor comments:

   _____

   _____

   _____

Lap welds:
Instructor comments:

_____
_____
_____

Butt welds:
Instructor comments:

_____
_____
_____

Fillet welds:
Instructor comments:

_____
_____
_____

3. On the fitted up joints from Work Order 17-3, weld the following:

Lap weld:
Instructor comments:

_____
_____
_____

Offset joint:
Instructor comments:

_____
_____
_____

Open butt joint:
Instructor comments:

_____
_____
_____

Butt joint with a backer:
Instructor comments:

_____
_____
_____

INSTRUCTOR COMMENTS:

_____
_____
_____
_____
_____
_____

Name _____ Date _____

Class _____ Instructor _____ Grade _____

1. *Technician A* says that safety glasses should be worn under a welder's helmet. *Technician B* says that a welder should protect his or her skin from the welding light, which can cause burns. Who is correct?
   A. Technician A only
   B. Technician B only
   C. Both Technicians A and B
   D. Neither Technician A nor B

2. *Technician A* says that bent structural parts must be replaced, not repaired. *Technician B* says that kinked structural parts must be replaced, not repaired. Who is correct?
   A. Technician A only
   B. Technician B only
   C. Both Technicians A and B
   D. Neither Technician A nor B

3. *Technician A* says that parts that are considered structural are the same on all vehicles. *Technician B* says that structural parts that are to be replaced should be cut off before rough pulling. Who is correct?
   A. Technician A only
   B. Technician B only
   C. Both Technicians A and B
   D. Neither Technician A nor B

4. *Technician A* says that the best way to identify a structural part is by looking it up in a repair manual. *Technician B* says that structural parts make a different sound than nonstructural parts when hit with a hammer. Who is correct?
   A. Technician A only
   B. Technician B only
   C. Both Technicians A and B
   D. Neither Technician A nor B

5. *Technician A* says that rough pulling is the initial pull and measuring is not necessary until after it is complete. *Technician B* says that measuring often and accurately is the key to a successful repair. Who is correct?
   A. Technician A only
   B. Technician B only
   C. Both Technicians A and B
   D. Neither Technician A nor B

6. *Technician A* says that safety is critical when working on structural pulling equipment and that work should be done safely and deliberately. *Technician B* says that when measuring a vehicle safety glasses and protective gloves are not necessary. Who is correct?
   A. Technician A only
   B. Technician B only
   C. Both Technicians A and B
   D. Neither Technician A nor B

7. *Technician A* says that making a repair plan is not necessary. *Technician B* says that even though a repair appears to be simple, even obvious, a repair plan forces the technician to precisely plan the procedure, thus making it more efficient. Who is correct?
   A. Technician A only
   B. Technician B only
   C. Both Technicians A and B
   D. Neither Technician A nor B

8. *Technician A* says that rough pulling may be necessary to accurately assess damage. *Technician B* says that kinked parts that will be removed for replacement often make good pulling locations for rough pulls. Who is correct?
   A. Technician A only
   B. Technician B only
   C. Both Technicians A and B
   D. Neither Technician A nor B

9. *Technician A* says that bent parts often crack when repairs are attempted. *Technician B* says that high-strength steel when pulled may develop small hard-to-identify cracks. Who is correct?
   A. Technician A only
   B. Technician B only
   C. Both Technicians A and B
   D. Neither Technician A nor B

10. *Technician A* says that butt joints should always have a backer. *Technician B* says that weld-through primer helps prevent corrosion. Who is correct?
    A. Technician A only
    B. Technician B only
    C. Both Technicians A and B
    D. Neither Technician A nor B

11. *Technician A* says that proper measuring during fit-up is critical. *Technician B* says that after taking a joint, it should be measured to confirm that it has not moved before continuing to weld. Who is correct?
    A. Technician A only
    B. Technician B only
    C. Both Technicians A and B
    D. Neither Technician A nor B

12. Technician A says that because of the heat from welding, parts should be measured often during the welding process to ensure accurate placement. Technician B says that when welding a vehicle, care should be taken to ensure that nearby parts are not harmed. Who is correct?
    A. Technician A only
    B. Technician B only
    C. Both Technicians A and B
    D. Neither Technician A nor B

13. **TRUE** or **FALSE**. A tack weld and a spot weld are the same.

14. *Technician A* says that a practice weld should be made on similar steel prior to welding a part to ensure a good weld. *Technician B* says that adhesive bonding is always recommended as a substitute for welding. Who is correct?
    A. Technician A only
    B. Technician B only
    C. Both Technicians A and B
    D. Neither Technician A nor B

15. *Technician A* says that when using adhesive bonding the parts must be held firmly in place until the adhesive has cured. *Technician B* says that the adhesive bonded parts can be held too tightly, thus squeezing out the needed bonding agent and making an insufficient joint. Who is correct?
    A. Technician A only
    B. Technician B only
    C. Both Technicians A and B
    D. Neither Technician A nor B

ASE-STYLE REVIEW QUESTIONS

1. *Technician A* says that all welded-on panels are structural panels. *Technician B* says that although welded-on panels are often structural parts, not all are. Who is correct?
   A. Technician A only
   B. Technician B only
   C. Both Technicians A and B
   D. Neither Technician A nor B

2. *Technician A* says that if a part is bent, it should be repaired. *Technician B* says that if a part is kinked, it should be replaced. Who is correct?
   A. Technician A only
   B. Technician B only
   C. Both Technicians A and B
   D. Neither Technician A nor B

3. *Technician A* says that the decision about when to replace a part can be based on the manufacturer's recommendation. *Technician B* says that the cost of repair is often used to judge whether a part should be replaced instead of repairing it. Who is correct?
   A. Technician A only
   B. Technician B only
   C. Both Technicians A and B
   D. Neither Technician A nor B

4. *Technician A* says that all welded replacement parts are installed at factory seams. *Technician B* says that sectioning cut lines are always left to the technician's discretion. Who is correct?
   A. Technician A only
   B. Technician B only
   C. Both Technicians A and B
   D. Neither Technician A nor B

5. *Technician A* says that crashworthiness is how a vehicle is designed to perform in a collision. *Technician B* says that crashworthiness is how a vehicle is designed to absorb energy during a collision. Who is correct?
   A. Technician A only
   B. Technician B only
   C. Both Technicians A and B
   D. Neither Technician A nor B

6. *Technician A* says that a lap joint is the type of joint used in sectioning all welded-on parts. *Technician B* says that a lap joint is often used to repair trunk floors. Who is correct?
   A. Technician A only
   B. Technician B only
   C. Both Technicians A and B
   D. Neither Technician A nor B

7. *Technician A* says that a butt joint can be used either with a backer or without. *Technician B* says that to make a strong butt weld, a backer must be used. Who is correct?
   A. Technician A only
   B. Technician B only
   C. Both Technicians A and B
   D. Neither Technician A nor B

8. *Technician A* says that spot welds are removed using a hole punch. *Technician B* says that an air chisel is the best tool for removing factory spot welds. Who is correct?
   A. Technician A only
   B. Technician B only
   C. Both Technicians A and B
   D. Neither Technician A nor B

9. *Technician A* says that the manufacturer recommendation for application of adhesive bond should be followed precisely. *Technician B* says that all adhesive bonding is mixed at the same ratio. Who is correct?
   A. Technician A only
   B. Technician B only
   C. Both Technicians A and B
   D. Neither Technician A nor B

10. *Technician A* says that measuring and fit-up are critical when replacing welded-on replacement parts. *Technician B* says that when welding on a replacement part, the technician should check the fit often because the act of welding may move the part slightly due to heat. Who is correct?
    A. Technician A only
    B. Technician B only
    C. Both Technicians A and B
    D. Neither Technician A nor B

11. *Technician A* says that all replacement door skins are adhesive bonded in place without the use of welds. *Technician B* says that all replacement door skins are welded in place without the use of adhesive bonding. Who is correct?
    A. Technician A only
    B. Technician B only
    C. Both Technicians A and B
    D. Neither Technician A nor B

12. *Technician A* says that a technician may use a combination of welds and adhesive bonding when replacing a quarter panel. *Technician B* says that before a quarter panel is replaced it should be measured for length, width, and height. Who is correct?
    A. Technician A only
    B. Technician B only
    C. Both Technicians A and B
    D. Neither Technician A nor B

13. *Technician A* says that side curtain air bags may accidentally deploy if not disabled when replacing a roof panel. *Technician B* says that the back light should be removed when replacing a roof panel. Who is correct?
    A. Technician A only
    B. Technician B only
    C. Both Technicians A and B
    D. Neither Technician A nor B

14. *Technician A* says that factory spot welds are easily found. *Technician B* says that weld-through primer is always used when welding on a replacement panel. Who is correct?
    A. Technician A only
    B. Technician B only
    C. Both Technicians A and B
    D. Neither Technician A nor B

ESSAY-STYLE REVIEW QUESTIONS

1. Describe what should be considered in the decision to replace vs. repair a part?

   _____

   _____

   _____

2. Explain what a kinked part is.

   _____

   _____

   _____

3. In your own words, explain crashworthiness.

   _____

   _____

   _____

4. Where would a technician find a recommended procedure for sectioning a replacement part?

   _____

   _____

   _____

5. What is the best tool for removing a factory spot weld? Why?

   _____

   _____

   _____

6. What is destructive testing of a weld?

   _____

   _____

   _____

LABORATORY ACTIVITIES

1. Using a door that has been removed from a vehicle, remove the outer door skin.
2. Measure and fit a new door skin to the door.
3. Set up and practice an open butt weld.
4. Set up and practice a butt weld with a backer.
5. Set up and practice a lap weld.
6. Set up and practice plug welds.

TOPIC-RELATED MATH QUESTIONS

1. If a factory quarter panel had 15 spot welds and the recommendation was to increase the factory welds by 33%, how many total welds should a technician put on the replacement part? (20)
2. A quarter panel is 54 inches long with a recommendation to place a spot weld every 1.5 inches. How many spot welds are needed? (36)
3. A 39-inch long replacement rail must have 13.5 inches removed. What will be the length of the remaining rail? (25.5)
4. A replacement roof is 54 inches long and 32.5 inches wide. One tube of adhesive will cover 76 inches. How many tubes are needed? (2.28)

CRITICAL THINKING CHALLENGES

1. The replacement part cost and the labor cost for its replacement would total $250. The cost of repair is $197.50. Should the part be replaced or repaired? Why?

   _____

   _____

   _____

2. Which is a better repair, welding or adhesive bonding? Why?

   _____

   _____

   _____

3. Which is a better repair, a sectioning or replacing at factory seams?

   _____

   _____

   _____

4. Why is crashworthiness important?

   _____

   _____

   _____

5. Why should a technician perform destructive testing on practice welds before welding in a replacement part?

   _____

   _____

   _____

Chapter 18

Plastic Repair

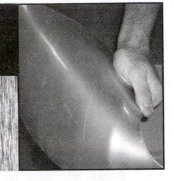

■ **WORK ASSIGNMENT 18-1**

PLASTIC IDENTIFICATION

Name _____   Date _____

Class _____   Instructor _____   Grade _____

NATEF SECTION II, PLASTICS AND ADHESIVES F. 1, 2

1. After reading the assignment, in the space provided below, list the personal and environmental safety equipment and precautions needed for this assignment.

_____

_____

_____

2. In the area below describe thermoset plastic.

_____

_____

_____

3. In the area below describe thermoplastic plastic.

_____

_____

_____

4. In the area below describe composite plastic.

_____

_____

_____

5. What is an ISO code and where can it be found?

_____

_____

_____

6. If the ISO code cannot be located on the part, describe alternative ways to identify the type of plastic a part is made of.

_____

_____

_____

7. Why is it important to know the type of plastic when repairing plastic?

_____

_____

_____

8. Why is it not possible to weld thermoset plastic?

_____

_____

_____

9. Why or why not is identifying a part as a polyolefin important?

_____

_____

_____

INSTRUCTOR COMMENTS:

_____

_____

_____

_____

_____

_____

PLASTIC TEST

Name _____   Date _____

Class _____   Instructor _____   Grade _____

NATEF SECTION II, PLASTICS AND ADHESIVES F. 1, 2

1. After reading the assignment, in the space provided below, list the personal and environmental safety equipment and precautions needed for this assignment.

   _____

   _____

   _____

2. Describe the float test for identifying plastic.

   _____

   _____

   _____

3. Describe the sand test for identifying plastic.

   _____

   _____

   _____

4. Describe the flexibility test for identifying plastic.

   _____

   _____

   _____

5. Explain what plastic memory is.

   _____

   _____

   _____

INSTRUCTOR COMMENTS:

_____

_____

_____

_____

_____

PLASTIC REPAIR

Name _____   Date _____

Class _____   Instructor _____   Grade _____

NATEF SECTION II, PLASTICS AND ADHESIVES F. 3, 4, 5

1. In the space provided below, list the personal and environmental safety precautions necessary for this assignment.

   _____

   _____

   _____

2. Explain the steps in the process of hot air plastic welding.

   _____

   _____

   _____

3. Explain the steps in the process of airless plastic welding.

   _____

   _____

   _____

4. Explain when a single-sided adhesive repair would be performed and when a two-sided repair would be performed.

   _____

   _____

   _____

5. List the steps used for a two-sided repair.

   _____

   _____

   _____

6. List the steps used for a single-sided repair.

   _____

   _____

   _____

7. What is an SMC plastic?

   _____

   _____

   _____

8. What is an FRP plastic?

   _____

   _____

   _____

9. Why is it important to determine the difference between SMC and FRP?

_____

_____

_____

INSTRUCTOR COMMENTS:

_____

_____

_____

_____

_____

_____

PLASTIC IDENTIFICATION

Name _____   Date _____

Class _____   Instructor _____   Grade _____

OBJECTIVES

- Know, understand, and use the safety equipment necessary for the task.
- Demonstrate the personal, shop, and environmental safety precautions that a technician must take when working with plastics.
- Distinguish which parts are either thermoset plastics or thermoplastics.
- Using ISO charts or a service manual, distinguish different types of plastics.
- Demonstrate how to identify the different types of plastics.

NATEF TASK CORRELATION

The written and hands-on activities in this chapter satisfy the NATEF High Priority-Individual and High Priority-Group requirements for Section II, Plastics and Adhesives F. 1, 2.

Tools and equipment needed (NATEF tool list)

- Safety glasses
- Gloves
- Ear protection
- Particle mask
- Pencil and paper
- Assorted hand tools
- ISO chart
- Vehicle service manuals

Vehicle Description

Year_____ Make _____ Model _____

VIN _____ Paint Code _____

PROCEDURE

1. After reading the work order, gather the safety gear needed to complete the task. In the space provided below, list the personal and environmental safety equipment and precautions needed for this assignment. Have the instructor check and approve your plan before proceeding.

_____

_____

_____

INSTRUCTOR'S APPROVAL _____

2. With the parts provided, determine if they are either a thermoset or a thermoplastic. Record your findings:

_____

_____

_____

3. With the parts provided, identify their ISO codes. Record your findings:.

_____

_____

_____

4. If a part's ISO code cannot be found, use its service manual to identify the type of plastic. Record your findings:

_____

_____

_____

5. With the parts provided, identify the parts that are polyolefin and those that are not. Record your findings:

_____

_____

_____

INSTRUCTOR COMMENTS:

_____

_____

_____

_____

_____

_____

PLASTIC TEST

Name _____ Date _____

Class _____ Instructor _____ Grade _____

OBJECTIVES

- Know, understand, and use the safety equipment necessary for the task.
- Demonstrate the processes for the plastic float test.
- Demonstrate the processes for the plastic sand test.
- Demonstrate the processes for the flexibility test.
- Demonstrate plastic repair using the plastics "memory."

NATEF TASK CORRELATION

The written and hands-on activities in this chapter satisfy the NATEF High Priority-Individual and High Priority-Group requirements for NATEF Section II, Plastics and Adhesives F. 1, 2.

Tools and equipment needed (NATEF tool list)

- Safety glasses
- Gloves
- Ear protection
- Particle mask
- Pencil and paper
- Assorted hand tools
- Containers of water
- DA sander
- Assorted DA sandpaper
- Razor blades

Vehicle Description

Year_____ Make _____ Model _____

VIN _____ Paint Code _____

PROCEDURE

1. After reading the work order, gather the safety gear needed to complete the task. In the space provided below, list the personal and environmental safety equipment and precautions needed for this assignment. Have the instructor check and approve your plan before proceeding.

_____

_____

_____

INSTRUCTOR'S APPROVAL _____

2. Using the plastic parts provided, identify the plastic classification by using the float test. Describe the process.

_____

_____

_____

Record your findings:

_____

_____

_____

3. Using the plastic parts provided, identify the plastic classification by using the sand test. Describe the process.

_____

_____

_____

Record your findings:

_____

_____

_____

4. Using the plastic parts provided, identify the plastic classification by using the flexibility test. Describe the process.

_____

_____

_____

Record your findings:

_____

_____

_____

5. Demonstrate plastic repair using heat and the plastics memory to complete your repair. Record your findings:

_____

_____

_____

INSTRUCTOR COMMENTS:

_____

_____

_____

_____

_____

_____

1. *Technician A* says that plastic parts can be identified by their location on the vehicle. *Technician B* says that all plastic is repaired with adhesive and it is not important to know which plastic it is. Who is correct?
   A. Technician A only
   B. Technician B only
   C. Both Technicians A and B
   D. Neither Technician A nor B

2. *Technician A* says that when repairing plastic the MSDS should be checked to determine which personal protective equipment is needed. *Technician B* says that a parts ISO code will identify the type of plastic used to manufacture the part. Who is correct?
   A. Technician A only
   B. Technician B only
   C. Both Technicians A and B
   D. Neither Technician A nor B

3. *Technician A* says that a thermoplastic will repeatedly soften without changing its makeup. *Technician B* says that a thermoplastic cannot be plastic welded. Who is correct?
   A. Technician A only
   B. Technician B only
   C. Both Technicians A and B
   D. Neither Technician A nor B

4. *Technician A* says that a thermoset will repeatedly soften without changing its makeup. *Technician B* says that a thermoset cannot be plastic welded. Who is correct?
   A. Technician A only
   B. Technician B only
   C. Both Technicians A and B
   D. Neither Technician A nor B

5. *Technician A* says that the float test is only conclusive if the sliver of plastic sinks. *Technician B* says that if a sliver of plastic dropped into water floats, this indicates that it is a thermoset. Who is correct?
   A. Technician A only
   B. Technician B only
   C. Both Technicians A and B
   D. Neither Technician A nor B

6. *Technician A* says that if plastic gets gummy, smears, or gets waxy during the sand test it is not a polyolefin. *Technician B* says that a polyolefin is a thermoset compound. Who is correct?
   A. Technician A only
   B. Technician B only
   C. Both Technicians A and B
   D. Neither Technician A nor B

7. *Technician A* says that it is important to identify plastic components to know which type of repair adhesive to use. *Technician B* says that some plastic repair systems identify parts by their flexibility instead of the compounds. Who is correct?
   A. Technician A only
   B. Technician B only
   C. Both Technicians A and B
   D. Neither Technician A nor B

8. *Technician A* says that a bumper that is damaged may return to its original shape when heated. *Technician B* says that when plastic becomes kinked it is hardened and should be replaced. Who is correct?
   A. Technician A only
   B. Technician B only
   C. Both Technicians A and B
   D. Neither Technician A nor B

9. *Technician A* says that if a plastic bumper is damaged but does not rip, a single-sided repair can be performed. *Technician B* says that if a plastic bumper is ripped it cannot be repaired. Who is correct?
   A. Technician A only
   B. Technician B only
   C. Both Technicians A and B
   D. Neither Technician A nor B

10. *Technician A* says that adhesives are commonly used to repair plastic parts for both single-sided and two-sided repairs. *Technician B* says a bumper that is ripped must be plastic welded. Who is correct?
    A. Technician A only
    B. Technician B only
    C. Both Technicians A and B
    D. Neither Technician A nor B

Chapter 19

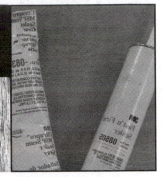

Corrosion Protection

■ WORK ASSIGNMENT 19-1

SAFETY AND TYPES OF CORROSION PROTECTION

Name _____   Date _____

Class _____   Instructor _____   Grade _____

NATEF TASKS I A. 9, 13; II B. 11, 13; IV B. 7, 20

1. After reading the assignment, in the space provided below, list the personal and environmental safety equipment and precautions needed for this assignment.

   _____

   _____

   _____

2. What is corrosion? Record your findings:

   _____

   _____

3. Why are some metals considered self-healing? Record your findings:

   _____

   _____

4. How does galvanization work and why? Record your findings:

   _____

   _____

5. What is E-coat and how is it applied? Record your findings:

   _____

   _____

6. How do seam sealers protect a vehicle from corrosion? Record your findings:

   _____

   _____

   _____

INSTRUCTOR COMMENTS:

_____

_____

_____

_____

_____

_____

CAUSES OF OXIDATION, CORROSIVE "HOT SPOTS," AND CORROSIVE COMPOUNDS

Name _____ Date _____

Class _____ Instructor _____ Grade _____

NATEF TASKS I A. 9, 13; II B. 11, 13; IV B. 7, 20

1. After reading the assignment, in the space provided below, list the personal and environmental safety equipment and precautions needed for this assignment.

   _____

   _____

   _____

2. What is a "corrosive hot spot," and what causes it? Record your findings:

   _____

   _____

   _____

3. How are corrosive hot spots created during:
   A collision:

   _____

   _____

   _____

   Collision repair:

   _____

   _____

   Atmospheric exposure:

   _____

   _____

   Galvanic corrosion:

   _____

   _____

4. What is a corrosion insulator, and how does it work? Record your findings:

   _____

   _____

5. Describe how each of the anti-corrosive compounds listed below works.
   Chemical etching:

   _____

   _____

   _____

Conversion coating:

_____

_____

_____

Galvanic corrosion:

_____

_____

_____

Basecoats/undercoats:

_____

_____

_____

Weld-through primer:

_____

_____

_____

Self-etching/wash primers:

_____

_____

_____

Epoxy primers:

_____

_____

_____

Petroleum based:

_____

_____

_____

Wax based:

_____

_____

Rubberized material:

_____

_____

_____

INSTRUCTOR COMMENTS:

_____

_____

_____

_____

_____

_____

APPLICATION OF ANTI-CORROSIVE COMPOUNDS

Name _____   Date _____

Class _____   Instructor _____   Grade _____

NATEF TASKS I A. 9, 13; II B. 11, 13; IV B. 7, 20

1. After reading the assignment, in the space provided below, list the personal and environmental safety equipment and precautions needed for this assignment.

   _____

   _____

   _____

2. Why is surface preparation for application of corrosion protection important? Record your findings:

   _____

   _____

   _____

3. Explain the items below for each of the corrosion protection materials listed.
   Application of self-etching primer:

   _____

   _____

   _____

   Its limitations and cautions:

   _____

   _____

   _____

   Preparation:

   _____

   _____

   _____

   Application:

   _____

   _____

   _____

   Other substances that should follow the corrosion protector:

   _____

   _____

   _____

   Application of epoxy primer:

   _____

   _____

   _____

Limitations/cautions:

_____
_____
_____

Preparation:

_____
_____
_____

Application:

_____
_____
_____

Other substances that should follow the corrosion protector:

_____
_____
_____

Application of seam sealers and foams:

_____
_____
_____

Limitations/cautions:

_____
_____
_____

Preparation:

_____
_____
_____

Application:

_____
_____
_____

Other substances that should follow the corrosion protector:

_____
_____
_____

INSTRUCTOR COMMENTS:

_____
_____
_____
_____
_____
_____

SAFETY AND TYPES OF CORROSION PROTECTION

Name _____   Date _____

Class _____   Instructor _____   Grade _____

OBJECTIVES

- Discuss the methods employed by the automobile manufacturing industry to prevent oxidation and corrosion.
- Define corrosion and understand the elements that promote formation and growth of oxidation.
- Identify environments that may promote oxidation on vehicle bodies and undercarriages.
- Define corrosion hot spots and explain how they impact the collision repair industry.
- Discuss metal treatment techniques to retard oxidation on automobiles.
- Discuss the types of anti-corrosion materials used to prevent or inhibit corrosion.
- Identify the corrosion protective basecoat materials commonly used in repairing collision-damaged vehicles.
- Identify various seam sealing materials and discuss their intended uses.

NATEF TASK CORRELATION

The written and hands-on activities in this chapter satisfy the NATEF High Priority-Individual and High Priority-Group requirements for Section I A. 9, 13; II B. 11, 13; IV B. 7, 20.

Tools and equipment needed (NATEF tool list)

- Safety glasses
- Ear protection
- Pencil and paper
- Damaged vehicle
- Vehicle service manual
- Disposable clean-up towels
- Spray gun
- Wax and petroleum anti-corrosive compounds with applicator gun
- Gloves
- Particle mask
- Assorted hand tools
- Estimate or work order for the vehicle
- Wax and grease remover
- Anti-corrosive compounds
- Foams and applicator gun

Vehicle Description

Year_____ Make _____ Model _____

VIN _____ Paint Code _____

PROCEDURE

1. After reading the assignment, in the space provided below, list the personal and environmental safety equipment and precautions needed for this assignment. Have the instructor review and approve your list.

_____

_____

_____

INSTRUCTOR'S APPROVAL _____

Using the vehicle and work order/estimate provided by your instructor, complete the tasks below.

2. On an exposed panel:

Prepare the surface for etch material and conversion coating.
Instructor comment:

_____

_____

_____

Apply (according to production sheet recommendations) etch material.
Instructor comment:

_____

_____

_____

Apply (according to production sheet recommendations) conversion coating.
Instructor comment:

_____

_____

_____

Apply (according to production sheet recommendations) basecoat/undercoat.
Instructor comment:

_____

_____

_____

INSTRUCTOR COMMENTS:

_____

_____

_____

_____

_____

_____

APPLICATION OF ANTI-CORROSIVE COMPOUNDS

Name _____ Date _____

Class _____ Instructor _____ Grade _____

OBJECTIVES

- Discuss the methods employed by the automobile manufacturing industry to prevent oxidation and corrosion.
- Define corrosion and understand the elements that promote formation and growth of oxidation.
- Identify environments that may promote oxidation on vehicle bodies and undercarriages.
- Define corrosion hot spots and explain how they impact the collision repair industry.
- Discuss metal treatment techniques to retard oxidation on automobiles.
- Discuss the types of anti-corrosion materials used to prevent or inhibit corrosion.
- Identify the corrosion protective basecoat materials commonly used in repairing collision-damaged vehicles.
- Identify various seam sealing materials and discuss their intended uses.

NATEF TASK CORRELATION

The written and hands-on activities in this chapter satisfy the NATEF High Priority-Individual and High Priority-Group requirements for Section I A. 9, 13; II B. 11, 13; IV B. 7, 20.

Tools and equipment needed (NATEF tool list)

- Safety glasses
- Ear protection
- Pencil and paper
- Damaged vehicle
- Vehicle service manual
- Disposable clean-up towels
- Spray gun
- Wax and petroleum anti-corrosive compounds with applicator gun
- Gloves
- Particle mask
- Assorted hand tools
- Estimate or work order for the vehicle
- Wax and grease remover
- Anti-corrosive compounds
- Foams and applicator gun

Vehicle Description

Year_____ Make _____ Model _____

VIN _____ Paint Code _____

PROCEDURE

1. After reading the assignment, in the space provided below, list the personal and environmental safety equipment and precautions needed for this assignment. Have the instructor review and approve your list.

_____

_____

_____

INSTRUCTOR'S APPROVAL _____

Using the vehicle and work order/estimate provided by your instructor, complete the tasks below.

2. On an enclosed panel:

   Prepare the surface for foam application.
   Instructor comment:

   _____

   _____

   _____

   Inject pillar foam (according to production sheet recommendations).
   Instructor comment:

   _____

   _____

   _____

3. On an enclosed panel:

   Prepare the surface for structural foam application.
   Instructor comment:

   _____

   _____

   _____

   Inject structural foam (according to production sheet recommendations).
   Instructor comment:

   _____

   _____

   _____

4. On an exposed seam:

   Prepare the surface for seam sealer application.
   Instructor comment:

   _____

   _____

   _____

   Apply (according to production sheet recommendations) seam sealer.
   Instructor comment:

   _____

   _____

   _____

INSTRUCTOR COMMENTS:

_____

_____

_____

_____

_____

_____

Name _____   Date _____

Class _____   Instructor _____   Grade _____

1. *Technician A* says that a corrosion hot spot can be the result of not coating pry pick marks on the backside after repairing a damaged panel. *Technician B* says that corrosion hot spots only occur when heat is used for repairing damage. Who is correct?
   A. Technician A only
   B. Technician B only
   C. Both Technicians A and B
   D. Neither Technician A nor B

2. Which of the following is NOT one of the corrosion protective steps that automobile manufacturers use?
   A. E-coating
   B. passivation
   C. galvanizing
   D. wet-on-wet cleansing and etching

3. *Technician A* says that a properly applied anti-corrosion compound to the interior of a door should be self-healing. *Technician B* says that before applying any corrosion protecting materials, one should follow several required steps for refinishing operations as well. Who is correct?
   A. Technician A only
   B. Technician B only
   C. Both Technicians A and B
   D. Neither Technician A nor B

4. All of the following are links for corrosion to occur EXCEPT:
   A. aluminum
   B. moisture
   C. oxygen
   D. heat

5. *Technician A* says that steel oxidizing is a form of passivation. *Technician B* says that galvanized steel is a form of sacrificial corrosion. Who is correct?
   A. Technician A only
   B. Technician B only
   C. Both Technicians A and B
   D. Neither Technician A nor B

6. Which of the following is the recommended procedure when plastic filler is part of the repair scenario?
   A. etching the metal prior to applying plastic filler
   B. spraying self-etch primer onto the surface prior to applying plastic filler
   C. applying anti-corrosion compound prior to applying plastic filler
   D. cleaning the metal and applying filler directly over it

7. *Technician A* says that zinc phosphate coating is nearly an airtight coating over the bare metal surface. *Technician B* says that zinc phosphate coating must be covered with a topcoat in order to preserve its corrosion-protection capabilities. Who is correct?
   A. Technician A only
   B. Technician B only
   C. Both Technicians A and B
   D. Neither Technician A nor B

8. Which of the following is best suited for corrosion protection on the inside of the doors?
   A. self-etch primer
   B. wash primer
   C. wax-based anti-corrosion compound
   D. petroleum-based anti-corrosion compound

9. *Technician A* says that the dwell time for phosphoric acid on bare metal is approximately 1 minute. *Technician B* says that it is advisable to scrub the metal surface with a scuffing pad after the phosphoric acid has dried from the surface to remove the excess coating. Who is correct?
   A. Technician A only
   B. Technician B only
   C. Both Technicians A and B
   D. Neither Technician A nor B

10. *Technician A* says that weld-through coating is designed to liquefy and reflow around the weld nugget to protect the weld joint. *Technician B* says that prior to applying two-part epoxy primer, the surface should be chemically etched to obtain optimum results. Who is correct?
    A. Technician A only
    B. Technician B only
    C. Both Technicians A and B
    D. Neither Technician A nor B

11. *Technician A* says that most corrosion hot spots can be slowed by etching, priming, and sealing them from the elements. *Technician B* says that some anti-corrosion compounds are made of recycled rubber. Who is correct?
    A. Technician A only
    B. Technician B only
    C. Both Technicians A and B
    D. Neither Technician A nor B

12. *Technician A* says that seam sealers can be applied over bare metal surfaces. *Technician B* says that aluminum creates a self-protecting anti-corrosion coating as it ages. Who is correct?
    A. Technician A only
    B. Technician B only
    C. Both Technicians A and B
    D. Neither Technician A nor B

13. *Technician A* says that the zinc coating on the galvanized steel corrodes to protect the steel surface. *Technician B* says that aluminum is more corrosion resistant than uncoated steel. Who is correct?
    A. Technician A only
    B. Technician B only
    C. Both Technicians A and B
    D. Neither Technician A nor B

14. *Technician A* says that galvanic corrosion occurs when two dissimilar metals come into contact with each other. *Technician B* says that an insulated steel bolt or one with a special coating can be used to hold two aluminum parts together. Who is correct?
    A. Technician A only
    B. Technician B only
    C. Both Technicians A and B
    D. Neither Technician A nor B

15. *Technician A* says that sulfur dioxide is one of the key elements that causes acid rain. *Technician B* says that a marine environment commonly results in an accelerated oxidation process. Who is correct?
    A. Technician A only
    B. Technician B only
    C. Both Technicians A and B
    D. Neither Technician A nor B

Chapter 20

Paint Spray Guns

■ WORK ASSIGNMENT 20-1

SPRAY GUN SAFETY

Name _____   Date _____

Class _____   Instructor _____   Grade _____

NATEF TASKS A: 1, 2, 3, 4, 5, & 6.

1. After reading the assignment, in the space provided below, list the personal and environmental safety equipment and precautions needed for this assignment.

_____

_____

_____

2. Explain the statement "all accidents can be prevented" and list some of the precautions that should be taken in the paint department.

_____

_____

_____

3. What are PPDs and what should you use when spraying basecoat?

_____

_____

_____

4. List some of the environmental safety precautions that should be observed in a paint department.

_____

_____

_____

5. How do spray guns affect the air we breathe?

_____

_____

_____

6. Will the reduction of material use in the paint department affect the environment? How?

_____

_____

_____

7. Why should you wear safety glasses when using a respirator?

_____

_____

_____

8. In the space provided below, list the personal and environmental safety precautions necessary when painting.

_____

_____

_____

INSTRUCTOR COMMENTS:

_____

_____

_____

_____

_____

_____

■ WORK ASSIGNMENT 20-2

SPRAY GUN TYPES AND DESIGN

Name _____   Date _____

Class _____   Instructor _____   Grade _____

NATEF SECTION IV. PAINTING AND REFINISHING C. 1

1. After reading the assignment, in the space provided below, list the personal and environmental safety equipment and precautions needed for this assignment.

   _____

   _____

   _____

2. Describe a siphon feed spray gun.

   _____

   _____

   _____

3. Describe a gravity feed spray gun.

   _____

   _____

   _____

4. Describe a pressure feed spray gun.

   _____

   _____

   _____

5. What is an HVLP spray gun?

   _____

   _____

   _____

6. How is powder coating applied?

   _____

   _____

   _____

7. What is CFM?

   _____

   _____

   _____

8. What is PSI?

   _____

   _____

   _____

INSTRUCTOR COMMENTS:

_____
_____
_____
_____
_____
_____

SPRAY GUN COMPONENTS AND AIR SUPPLY

Name _____ Date _____

Class _____ Instructor _____ Grade _____

NATEF SECTION IV. PAINTING AND REFINISHING C. 2, 3

1. After reading the assignment, in the space provided below, list the personal and environmental safety equipment and precautions needed for this assignment.

    _____

    _____

    _____

2. What is the function of a spray gun's air cap?

    _____

    _____

    _____

3. What function does the auxiliary air cap nozzle produce when spraying paint?

    _____

    _____

    _____

4. Explain how atomization is accomplished with a gravity feed spray gun.

    Stage I:

    _____

    _____

    _____

    Stage II:

    _____

    _____

    _____

    Stage III:

    _____

    _____

5. What is and what function does a fluid needle perform?

    _____

    _____

    _____

6. What is and what function does an air cap perform?

    _____

    _____

    _____

7. What is and what function does the fluid control knob perform?

    _____

    _____

    _____

8. What is and what function does the pattern control knob perform?

_____

_____

_____

9. Why is it critical to use high-volume quick connect hose fittings with an HVLP gun?

_____

_____

_____

10. How does the hose size affect the operations of an HVLP spray gun?

_____

_____

_____

INSTRUCTOR COMMENTS:

_____

_____

_____

_____

_____

_____

■ WORK ASSIGNMENT 20-4

GUN ADJUSTMENT AND CLEANING

Name _____ Date _____

Class _____ Instructor _____ Grade _____

NATEF SECTION IV. PAINTING AND REFINISHING C. 2

1. After reading the assignment, in the space provided below, list the personal and environmental safety equipment and precautions needed for this assignment.

   _____

   _____

   _____

2. Of the three adjustments on a spray gun (air, fluid, and fan), which should be adjusted first and why?

   _____

   _____

   _____

3. Describe how you would find the recommended air pressure for a particular gun and the coating being sprayed.

   _____

   _____

   _____

4. Describe how you would adjust the air pressure for the gun provided.

   _____

   _____

   _____

5. Where is the fluid control knob located on the gun provided and what function does it serve?

   _____

   _____

   _____

6. Describe how you would adjust the fluid control knob on the gun provided.

   _____

   _____

   _____

7. Where is the pattern control knob located on the gun provided and what function does it serve?

   _____

   _____

   _____

8. Describe how you would adjust the fan control knob on the gun provided.

   _____

   _____

   _____

9. After the gun's air pressure, fluid, and fan have been adjusted, why should the air pressure be checked again?

_____

_____

_____

10. Describe cleaning of a spray gun.

_____

_____

_____

11. How would you clean the small orifices that have become clogged in the air cap?

_____

_____

_____

12. With what and where would you lubricate a spray gun?

_____

_____

_____

13. Why would a spray gun need rebuilding?

_____

_____

_____

INSTRUCTOR COMMENTS:

_____

_____

_____

_____

_____

_____

■ WORK ORDER 20-1

SPRAY GUN SAFETY

Name _____   Date _____

Class _____   Instructor _____   Grade _____

OBJECTIVES

- Know, understand, and use the safety equipment necessary for the task.
- Operate a spray gun using proper:
  - Personal safety
  - Environmental safety
  - Fellow worker protection safety

NATEF TASK CORRELATION

The written and hands-on activities in this chapter satisfy the NATEF High Priority-Individual and High Priority-Group requirements for Section IV, Painting and Refinishing, A. 1-6.

Tools and equipment needed (NATEF tool list)

- Safety glasses
- Ear protection
- Respirator
- Spray suit
- Pencil and paper
- Gloves
- Particle mask
- Air supply respirator
- Spray guns
- Assorted hand tools

Vehicle Description

Year_____ Make _____ Model _____

VIN _____ Paint Code _____

PROCEDURE

1. After reading the work order, gather the safety gear needed to complete the task. In the space provided below, list the personal and environmental safety equipment and precautions needed for this assignment. Have the instructor check and approve your plan before proceeding.

_____

_____

_____

INSTRUCTOR'S APPROVAL _____

2. Inspect the lab and identify any potential safety concerns. Record your findings:

_____

_____

3. With the use of its MSDS, find the required personal protective devices needed to spray primer filler. Record your findings:

_____

_____

_____

4. With the use of its MSDS, find the required personal protective devices needed to spray epoxy primer. Record your findings:

_____

_____

_____

5. With the use of its MSDS, find the required personal protective devices needed to spray basecoat. Record your findings:

_____

_____

_____

6. With the use of its MSDS, find the required personal protective devices needed to spray clearcoat. Record your findings:

_____

_____

7. Demonstrate the correct way to put on a respirator. Fit test both positive and negative.

INSTRUCTOR'S APPROVAL _____

8. Demonstrate the correct way to put on an air supply respirator.

INSTRUCTOR'S APPROVAL _____

9. Demonstrate the correct way to put on a particle mask.

INSTRUCTOR'S APPROVAL _____

INSTRUCTOR COMMENTS:

_____

_____

_____

_____

_____

_____

SPRAY GUN COMPONENTS AND AIR SUPPLY

Name _____   Date _____

Class _____   Instructor _____   Grade _____

OBJECTIVES

- Know, understand, and use the safety equipment necessary for the task.
- Inspect the components of a spray gun in preparation for use.
- Demonstrate the ability to properly choose the correct air supply for the equipment being used.
- Properly set regulator pressure.
- List proper air supply and regulation needs for refinish equipment.

NATEF TASK CORRELATION

The written and hands-on activities in this chapter satisfy the NATEF High Priority-Individual and High Priority-Group requirements for Section IV, Painting and Refinishing, C. 2, 3.

Tools and equipment needed (NATEF tool list)

- Safety glasses
- Ear protection
- Pencil and paper
- Air hose
- Regulated air supply
- Gloves
- Particle mask
- Assorted hand tools
- Spray guns

Gun Description

Type _____Make_____

Model _____Tip size _____

Coating to be sprayed _____ _____

PROCEDURE

1. After reading the work order, gather the safety gear needed to complete the task. In the space provided below, list the personal and environmental safety equipment and precautions needed for this assignment. Have the instructor check and approve your plan before proceeding.

_____

_____

_____

INSTRUCTOR'S APPROVAL _____

2. Remove, inspect, and clean, if necessary, the gun's air cap. Record your findings:

_____

_____

3. Were the auxiliary air cap nozzles clean? How did you or would you clean them? Record your findings:

_____

_____

4. Remove the spray gun's nozzle and needle. Inspect and clean if necessary. Record your findings:

_____

_____

_____

5. Choose from the air hoses supplied the correct one for use with an HVLP spray gun. What makes this hose correct?

_____

_____

_____

6. From the quick connect couplers supplied, choose the correct one to use with an HVLP gun. What makes this suppler correct?

_____

_____

_____

7. From the gun regulators supplied, choose the high flow one. What makes your choice correct?

_____

_____

_____

INSTRUCTOR COMMENTS:

_____

_____

_____

_____

_____

_____

_____

GUN ADJUSTMENT AND CLEANING

Name _____   Date _____

Class _____   Instructor _____   Grade _____

OBJECTIVES

- Know, understand, and use the safety equipment necessary for the task.
- Adjust the air pressure correctly.
- Adjust the fluid knob for best atomization.
- Adjust the fan pattern for application.
- Clean the spray gun.
- Lubricate the spray gun.

NATEF TASK CORRELATION

The written and hands-on activities in this chapter satisfy the NATEF High Priority-Individual and High Priority-Group requirements for Section IV, Painting and Refinishing, C. 2.

Tools and equipment needed (NATEF tool list)

- Safety glasses
- Ear protection
- Pencil and paper
- Spray guns
- Gun lubricant
- Gloves
- Particle mask
- Assorted hand tools
- Gun cleaning equipment
- Cleaning brushes

Gun Description

Type _____Make_____

Model _____Tip size _____

Coating to be sprayed _____

PROCEDURE

1. After reading the work order, gather the safety gear needed to complete the task. In the space provided below, list the personal and environmental safety equipment and precautions needed for this assignment. Have the instructor check and approve your plan before proceeding.

_____

_____

_____

INSTRUCTOR'S APPROVAL _____

2. From the gun regulators provided, choose a high-flow regulator. Why did you choose this one?

_____

_____

3. From the air hoses provided, choose the best one for an HVLP gun. Why did you choose this one?

_____

_____

4. Attach the high-flow regulator and correct hose and adjust the recommended air pressure. Describe your process.

_____

_____

_____

5. Find the recommended air pressure for the gun and coating being sprayed. Describe your process.

_____

_____

6. Set the proper fluid knob for best atomization. Describe your process.

_____

_____

_____

7. Set the fan for the object being painted. Describe your process.

_____

_____

_____

8. Demonstrate the proper cleaning of a gun. Describe your process.

_____

_____

_____

9. Demonstrate the lubrication of a gun. Describe your process.

_____

_____

_____

INSTRUCTOR COMMENTS:

_____

_____

_____

_____

_____

_____

Name _____ Date _____

Class _____ Instructor _____ Grade _____

1. *Technician A* says that if you spray in a paint booth with good airflow it is not critical to wear a respirator. *Technician B* says that if sanding in a prep deck that has good airflow it is not critical to wear a particle mask. Who is correct?
   A. Technician A only
   B. Technician B only
   C. Both Technicians A and B
   D. Neither Technician A nor B

2. *Technician A* says that all accidents can be prevented if safety is always considered a top priority. *Technician B* says that personal protective devices (PPDs) are the painter's first line of defense for safety. Who is correct?
   A. Technician A only
   B. Technician B only
   C. Both Technicians A and B
   D. Neither Technician A nor B

3. *Technician A* says that HVLP spray guns produce less overspray than non-HVLP guns. *Technician B* says that volatile organic compounds are harmless gasses released during paint application. Who is correct?
   A. Technician A only
   B. Technician B only
   C. Both Technicians A and B
   D. Neither Technician A nor B

4. *Technician A* says that the use of HVLP guns will increase the use of material. *Technician B* says that HVLP guns reduce overspray, VOC emission, and material use. Who is correct?
   A. Technician A only
   B. Technician B only
   C. Both Technicians A and B
   D. Neither Technician A nor B

5. *Technician A* says that safety glasses should be worn when sanding. *Technician B* says that safety glasses should be worn when using a respirator. Who is correct?
   A. Technician A only
   B. Technician B only
   C. Both Technicians A and B
   D. Neither Technician A nor B

6. *Technician A* says that PSI stands for personal safety initiative. *Technician B* says that PSI stands for pounds per square inch. Who is correct?
   A. Technician A only
   B. Technician B only
   C. Both Technicians A and B
   D. Neither Technician A nor B

7. *Technician A* says that CFM stands for cubic feet per minute. *Technician B* says that CFM is a measurement of air pressure. Who is correct?
   A. Technician A only
   B. Technician B only
   C. Both Technicians A and B
   D. Neither Technician A nor B

8. *Technician A* says that a gravity feed gun has the paint container on the bottom of the gun. *Technician B* says that a gravity feed gun only uses air pressure to move the paint out of the gun. Who is correct?
   A. Technician A only
   B. Technician B only
   C. Both Technicians A and B
   D. Neither Technician A nor B

9. *Technician A* says that powder coating is not a method of automotive paint application. *Technician B* says that rotating bell paint application is done in large collision repair shops. Who is correct?
   A. Technician A only
   B. Technician B only
   C. Both Technicians A and B
   D. Neither Technician A nor B

10. *Technician A* says that PSI is a measurement of pressure. *Technician B* says that HVPL guns use less materials than non-HVLP guns. Who is correct?
    A. Technician A only
    B. Technician B only
    C. Both Technicians A and B
    D. Neither Technician A nor B

11. *Technician A* says that air hoses, if they have enough pressure, will supply all the needed volume needed for an HVLP regardless of diameter. *Technician B* says that a 3/8-inch diameter is the minimum size air hose for the volume needed by an HVLP gun. Who is correct?
    A. Technician A only
    B. Technician B only
    C. Both Technicians A and B
    D. Neither Technician A nor B

12. *Technician A* says that the car cap orifices supply air pressure, which atomizes paint as it comes from the gun. *Technician B* says all air cap orifice air pressure forms the spray pattern. Who is correct?
    A. Technician A only
    B. Technician B only
    C. Both Technicians A and B
    D. Neither Technician A nor B

13. *Technician A* says that the spray gun's needle produces atomization. *Technician B* says that the fluid nozzle regulates the air pressure on the gun. Who is correct?
    A. Technician A only
    B. Technician B only
    C. Both Technicians A and B
    D. Neither Technician A nor B

14. *Technician A* says that the fluid control knob controls the amount of paint that comes out of the gun. *Technician B* says that the pattern (fan) control knob controls the amount of paint that comes out of the gun. Who is correct?
    A. Technician A only
    B. Technician B only
    C. Both Technicians A and B
    D. Neither Technician A nor B

15. *Technician A* says that paint atomization occurs outside the gun. *Technician B* says that the spray pattern shape starts to form in stage II of atomization. Who is correct?
    A. Technician A only
    B. Technician B only
    C. Both Technicians A and B
    D. Neither Technician A nor B

16. *Technician A* says that when adjusting a spray gun, air pressure should be adjusted first. *Technician B* says that when adjusting a spray gun, air pressure should be checked after adjusting the fan and fluid knobs. Who is correct?
    A. Technician A only
    B. Technician B only
    C. Both Technicians A and B
    D. Neither Technician A nor B

17. *Technician A* says that the fluid control knob regulates the amount of air that comes out of the gun. *Technician B* says that the fluid control knob adjusts the fan of the spray gun. Who is correct?
    A. Technician A only
    B. Technician B only
    C. Both Technicians A and B
    D. Neither Technician A nor B

18. *Technician A* says that the pattern adjustment knob adjusts the fan. *Technician B* says that the air pressure is measured in CFM. Who is correct?
    A. Technician A only
    B. Technician B only
    C. Both Technicians A and B
    D. Neither Technician A nor B

19. *Technician A* says that spray guns must be cleaned after each use. Technician B says that the spray gun's orifice should be cleaned with welding wire. Who is correct?
    A. Technician A only
    B. Technician B only
    C. Both Technicians A and B
    D. Neither Technician A nor B

20. *Technician A* says that spray guns should be lubricated. *Technician B* says that spray gun lubricant causes fisheye. Who is correct?
    A. Technician A only
    B. Technician B only
    C. Both Technicians A and B
    D. Neither Technician A nor B

Chapter 21

Spray Techniques

■ WORK ASSIGNMENT 21-1

ADJUSTING A SPRAY GUN

Name _____ Date _____

Class _____ Instructor _____ Grade _____

NATEF SECTION IV PAINTING AND REFINISHING SECTION C. 1, 2, 3.

1. After reading the assignment, in the space provided below, list the personal and environmental safety equipment and precautions needed for this assignment.

   _____

   _____

   _____

2. Describe the importance of adjusting a spray gun.

   _____

   _____

   _____

3. How do the items listed below affect finish?
   Travel speed:

   _____

   _____

   Distance:

   _____

   _____

   _____

4. Explain how regulator pressure, inlet pressure, and air cap pressure vary.

   _____

   _____

   _____

5. What does the fluid adjustment do and how should it be set?

   _____

   _____

   _____

6. What is atomization and how do air pressure and fluid adjustment affect it?

_____

_____

_____

7. What does the fan adjustment do and how is it set?

_____

_____

_____

8. Why should the air pressure be checked after the fluid and fan have been adjusted?

_____

_____

_____

9. How is texture affected by gun adjustment?

_____

_____

_____

10. What is and why is spray width important?

_____

_____

_____

11. How does the distance affect paint appearance?

_____

_____

_____

12. What is meant by stroke overlap and how does it affect the outcome of paint coverage?

_____

_____

_____

INSTRUCTOR COMMENTS:

_____

_____

_____

_____

_____

_____

PLAN OF ATTACK (SPRAY SEQUENCE)

Name _____ Date _____

Class _____ Instructor _____ Grade _____

NATEF SECTION IV PAINTING AND REFINISHING SECTION: C. 1, 2, 3; D: 3, 4, 5, 6

1. After reading the assignment, in the space provided below, list the personal and environmental safety equipment and precautions needed for this assignment.

   _____

   _____

   _____

2. What is and why is spray sequence (plan of attack) important?

   _____

   _____

   _____

3. How does the direction of air movement affect the plan of attack?

   _____

   _____

   _____

4. What is the "wet line" and why is it important?

   _____

   _____

   _____

5. When a technician tacks off the vehicle just prior to paint application, what should he or she be doing along with tacking the vehicle?

   _____

   _____

   _____

6. What is meant by "perpendicular" to the surface and why is it important?

   _____

   _____

   _____

7. Explain the terms below and why they are important.
   Distance:

   _____

   _____

   Arching:

   _____

   _____

   _____

Lead and lag:

_____

_____

_____

8. Why is triggering a gun important?

_____

_____

_____

INSTRUCTOR COMMENTS:

_____

_____

_____

_____

_____

_____

SPRAY TECHNIQUES

Name _____   Date _____

Class _____   Instructor _____   Grade _____

NATEF SECTION IV PAINTING AND REFINISHING SECTION: C. 1, 2, 3; D. 2, 3, 4, 5, 6

1. After reading the assignment, in the space provided below, list the personal and environmental safety equipment and precautions needed for this assignment.

   _____

   _____

   _____

2. Describe the process of cutting in parts (edging).

   _____

   _____

   _____

3. List the steps for painting a new E-coated part.

   _____

   _____

   _____

4. Describe the steps to apply single stage K2 paint.

   _____

   _____

   _____

5. How does the application of single stage solid colors differ from single stage metallic application?

   _____

   _____

   _____

6. How can streaking be controlled in single stage metallic paint?

   _____

   _____

   _____

7. How is basecoat paint applied?

   _____

   _____

   _____

8. How does fan width affect streaking during basecoat application?

   _____

   _____

   _____

9. How does overlap affect streaking during application?

   _____

   _____

10. How is clearcoat applied?

_____

_____

_____

11. How do panel painting and blending differ?

_____

_____

_____

INSTRUCTOR COMMENTS:

_____

_____

_____

_____

_____

_____

ADJUSTING A SPRAY GUN

Name _____   Date _____

Class _____   Instructor _____   Grade _____

OBJECTIVES

- Know, understand, and use the safety equipment necessary for the task.
- Demonstrate how to adjust a spray gun for proper:
  - Atomization
  - Fluid
  - Spray width
  - Distance
- Explain the proper stroke movement overlap and rate of movement for proper spraying.

NATEF TASK CORRELATION

The written and hands-on activities in this chapter satisfy the NATEF High Priority-Individual and High Priority-Group requirements for Section IV, Painting and Refinishing section C. 1, 2, 3.

Tools and equipment needed (NATEF tool list)

- Safety glasses
- Ear protection
- Respirator
- Spray suit
- Production sheets for paint
- Air pressure gauges
- Laser practice gun (optional)
- Assorted hand tools
- Gloves
- Particle mask
- Air supply respirator
- Spray guns
- MSDS for paint
- Air pressure regulators
- Pencil and paper

Vehicle Description

Year_____ Make _____ Model _____

VIN _____ Paint Code _____

Gun Description

Type _____Make_____

Model _____Tip size _____

Coating to be sprayed _____

PROCEDURE

1. Gather the safety gear needed to complete the task. In the space provided below, list the personal and environmental safety equipment needed and the precautions necessary for this assignment. Have the instructor check and approve your plan before proceeding.

_____

_____

_____

INSTRUCTOR'S APPROVAL _____

2. Pick a spray gun with a 1.4 needle and nozzle set-up.
   Gun's description:

_____

_____

3. How did you determine its set-up size?

_____

_____

4. Where would you find the recommendations for the paint coating provided?

_____

_____

5. Adjust the spray gun to its recommended air pressure. List your steps.

_____

_____

6. Adjust the spray width. List your steps.

_____

_____

7. Test the gun and fine-tune your adjustments. What was required?

_____

_____

8. Demonstrate proper overlap.

INSTRUCTOR COMMENTS:

_____

_____

_____

_____

_____

_____

PLAN OF ATTACK (SPRAY SEQUENCE)

Name _____   Date _____

Class _____   Instructor _____   Grade _____

OBJECTIVES

- Know, understand, and use the safety equipment necessary for the task.
- Demonstrate the proper spray sequence, including the plan of attack for large, small, tall, and short vehicles.
- Demonstrate the proper spray gun positions.

NATEF TASK CORRELATION

The written and hands-on activities in this chapter satisfy the NATEF High Priority-Individual and High Priority-Group requirements for Section IV, Painting and Refinishing section: C. 1, 2, 3; D. 3, 4, 5, 6.

Tools and equipment needed (NATEF tool list)

- Safety glasses
- Ear protection
- Respirator
- Spray suit
- Practice paint
- Pencil and paper
- Gloves
- Particle mask
- Air supply respirator
- Spray guns
- Practice paper and easel
- Assorted hand tools

Vehicle Description

Year_____ Make _____ Model _____

VIN _____ Paint Code _____

Gun Description

Type _____Make_____

Model _____Tip size _____

Coating to be sprayed _____

PROCEDURE

1. After reading the work order, gather the safety gear needed to complete the task. In the space provided below, list the personal and environmental safety equipment and precautions needed for this assignment. Have the instructor check and approve your plan before proceeding.

   _____
   _____
   _____

INSTRUCTOR'S APPROVAL _____

2. Using the vehicle provided, make a "plan of attack" (spray sequence). List your conclusions.

   _____

   _____

   _____

3. Using a second vehicle, tack it for paint application. During the process of tacking both, inspect it for readiness and make a plan of attack. List your conclusions.

   _____

   _____

   _____

4. In your paint booth, which direction should you spray? List your conclusions.

   _____

   _____

   _____

5. Spray a practice hood with the coating provided.
   Gun adjustment: _____

   Distance: _____

   Arching: _____

   Lead: _____

   Lag: _____

   Triggering: _____

INSTRUCTOR COMMENTS:

_____

_____

_____

_____

_____

_____

SPRAY TECHNIQUES

Name _____ Date _____

Class _____ Instructor _____ Grade _____

OBJECTIVES

- Know, understand, and use the safety equipment necessary for the task.
- Demonstrate the steps for different spray techniques, including:
  - Cutting in
  - Parts painting
  - Single stage sold color
  - Single stage metallic
  - Basecoat
  - Clearcoats
  - Panel painting

NATEF TASK CORRELATION

The written and hands-on activities in this chapter satisfy the NATEF High Priority-Individual and High Priority-Group requirements for Section IV, Painting and Refinishing section: C. 1, 2, 3; D. 2, 3, 4, 5, 6.

Tools and equipment needed (NATEF tool list)

- Safety glasses
- Ear protection
- Respirator
- Spray suit
- Single stage paint
- Production sheets
- Masking materials
- Assorted hand tools
- Gloves
- Particle mask
- Air supply respirator
- Spray guns
- Basecoat paint
- MSDS
- Pencil and paper

Vehicle Description

Year_____ Make _____ Model _____

VIN _____ Paint Code _____

Gun Description

Type _____Make_____

Model _____Tip size _____

Coating to be sprayed _____

PROCEDURE

1. After reading the work order, gather the safety gear needed to complete the task. In the space provided below, list the personal and environmental safety equipment and precautions needed for this assignment. Have the instructor check and approve your plan before proceeding.

_____

_____

_____

INSTRUCTOR'S APPROVAL _____

2. On the part provided cut it in (edge).

   Gun adjustment:

   _____

   _____

   Plan of attack:

   _____

   _____

   Distance:

   _____

   _____

   Overlap:

   _____

   _____

   Lead and lag:

   _____

   _____

   Final appearance:

   _____

   _____

3. Spray single stage paint.

   Gun adjustment:

   _____

   _____

   Plan of attack:

   _____

   _____

   Distance:

   _____

   _____

Overlap:

_____

_____

_____

Lead and lag:

_____

_____

_____

Final appearance:

_____

_____

_____

4. Spray basecoat.
   Gun adjustment:

_____

_____

_____

Plan of attack:

_____

_____

_____

Distance:

_____

_____

_____

Overlap:

_____

_____

_____

Lead and lag:

_____

_____

_____

Final appearance:

_____

_____

_____

5. Spray clear coat.
   Gun adjustment:

_____

_____

_____

Plan of attack:

_____

_____

_____

Distance:

_____

_____

_____

Overlap:

_____

_____

_____

Lead and lag:

_____

_____

_____

Final appearance:

_____

_____

_____

INSTRUCTOR COMMENTS:

_____

_____

_____

_____

_____

_____

Name _____   Date _____

Class _____   Instructor _____   Grade _____

1. *Technician A* says that a spray suit will protect the paint finish from contamination. *Technician B* says that a paint suit will protect the technician from contamination. Who is correct?
   A. Technician A only
   B. Technician B only
   C. Both Technicians A and B
   D. Neither Technician A nor B

2. *Technician A* says that adjusting a spray gun using the wall regulator gauge is sufficient for accurate spray gun pressure settings. *Technician B* says that adjusting the most accurate place to measure an HVLP gun pressure is at the cap. Who is correct?
   A. Technician A only
   B. Technician B only
   C. Both Technicians A and B
   D. Neither Technician A nor B

3. *Technician A* says that too slow a travel speed may cause runs. *Technician B* says that too fast a travel speed may cause a rough dry finish. Who is correct?
   A. Technician A only
   B. Technician B only
   C. Both Technicians A and B
   D. Neither Technician A nor B

4. *Technician A* says that distance from the surface has little effect on the finish. *Technician B* says that distance has little effect on transfer efficiency. Who is correct?
   A. Technician A only
   B. Technician B only
   C. Both Technicians A and B
   D. Neither Technician A nor B

5. Technician A says that the fluid adjustment knobs adjust how far back the fluid needle will travel. *Technician B* says that a proper fluid needle and nozzle set-up is critical for good coating application. Who is correct?
   A. Technician A only
   B. Technician B only
   C. Both Technicians A and B
   D. Neither Technician A nor B

6. *Technician A* says that fan width has little effect on overspray. *Technician B* says that texture is controlled by viscosity, not spray technique. Who is correct?
   A. Technician A only
   B. Technician B only
   C. Both Technicians A and B
   D. Neither Technician A nor B

7. *Technician A* says that in a down draft booth the technician should spray from the bottom of the vehicle upward. *Technician B* says that in a semi-down draft booth the technician should spray from the rear to the front. Who is correct?
   A. Technician A only
   B. Technician B only
   C. Both Technicians A and B
   D. Neither Technician A nor B

8. *Technician A* says that a plan of attack helps the spray technician be more efficient. *Technician B* says that when spraying clear, keeping good overlap and a consent wet line are important. Who is correct?
   A. Technician A only
   B. Technician B only
   C. Both Technicians A and B
   D. Neither Technician A nor B

9. *Technician A* says that final inspection and tacking can be done together. *Technician B* says that tacking will eliminate fingerprint marks. Who is correct?
   A. Technician A only
   B. Technician B only
   C. Both Technicians A and B
   D. Neither Technician A nor B

10. *Technician A* says that cutting in is spraying the edges of a new part before it is mounted on the vehicle. *Technician B* says that the apprentice painter often is given the task of edging in parts. Who is correct?
    A. Technician A only
    B. Technician B only
    C. Both Technicians A and B
    D. Neither Technician A nor B

11. *Technician A* says that if overlap is too wide (less than 50%) dry spray is controlled. *Technician B* says that if overlap is too wide (less than 50%) streaking may occur. Who is correct?
    A. Technician A only
    B. Technician B only
    C. Both Technicians A and B
    D. Neither Technician A nor B

12. *Technician A* says that streaking is more likely to occur in a metallic color than in a solid one. *Technician B* says that streaking does not occur in solid colors. Who is correct?
    A. Technician A only
    B. Technician B only
    C. Both Technicians A and B
    D. Neither Technician A nor B

13. *Technician A* says that blending helps color match. *Technician B* says that there are different methods of blending that help with particularly difficult colors. Who is correct?
    A. Technician A only
    B. Technician B only
    C. Both Technicians A and B
    D. Neither Technician A nor B

14. *Technician A* says the wet bedding helps to control difficult metallic colors. *Technician B* says that standard blending helps control difficult metallic colors. Who is correct?
    A. Technician A only
    B. Technician B only
    C. Both Technicians A and B
    D. Neither Technician A nor B

15. *Technician A* says that standard blending can be used with solid colors. *Technician B* says that clearcoat provides durability and gloss. Who is correct?
    A. Technician A only
    B. Technician B only
    C. Both Technicians A and B
    D. Neither Technician A nor B

Chapter 22

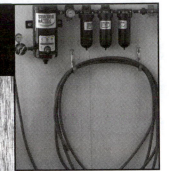

Refinishing Shop Equipment

■ WORK ASSIGNMENT 22-1

COMPRESSED AIR AND ITS DELIVERY

Name _____   Date _____

Class _____   Instructor _____   Grade _____

This section identifies specific paint department tools and their use and maintenance.

1. After reading the assignment, in the space provided below, list the personal and environmental safety equipment and precautions needed for this assignment.

   _____

   _____

   _____

2. Describe how a piston compressor operates.

   _____

   _____

3. How does a two-stage piston compressor work?

   _____

   _____

4. What is a rotary screw compressor and how does it work?

   _____

   _____

5. What special considerations must be taken to produce breathable air for an air supply respirator?

   _____

   _____

6. How would a shop calculate the amount of air that would be needed to supply a specific shop?

   _____

   _____

7. What is meant by CFM and what does it measure?

   _____

   _____

8. What is meant by PSI and what does it measure?

_____

_____

_____

9. What is meant by a 55% duty cycle compressor?

_____

_____

_____

10. How does a supply hose size affect an HVLP spray gun?

_____

_____

_____

11. What is an air transformer?

_____

_____

_____

12. What does an air extractor do?

_____

_____

_____

13. What does a refrigerant evaporator do in a compressed air system?

_____

_____

_____

INSTRUCTOR COMMENTS:

_____

_____

_____

_____

_____

_____

■ WORK ASSIGNMENT 22-2

SUPPLY HOSES AND COMPRESSOR MAINTENANCE

Name _____   Date _____

Class _____   Instructor _____   Grade _____

This section identifies specific paint department tools and their use and maintenance.

1. After reading the assignment, in the space provided below, list the personal and environmental safety equipment and precautions needed for this assignment.

   _____
   _____
   _____

2. How does a single braided hose differ from a double braided hose?

   _____
   _____
   _____

3. How does heat affect a spray booth hose?

   _____
   _____
   _____

4. How is CFM capacity affected by the supply hose ID?

   _____
   _____
   _____

5. A spray technician is given a supply hose of 3/8 ID and 100 ft long. Approximately how much air pressure will be lost? Why?

   _____
   _____
   _____

6. Why should a paint technician choose a high-flow quick connector for the paint department?

   _____
   _____
   _____

7. What is hose delaminating?

   _____
   _____
   _____

8. List the preventive maintenance steps that a technician should perform on a compressor daily.

   _____
   _____
   _____

9. Why should a supply hose connector be facing downward?

_____

_____

_____

10. Why is preventive maintenance important with air compressors?

_____

_____

_____

11. List the preventive maintenance steps that a technician should perform on a compressor daily.

_____

_____

_____

12. List the preventive maintenance steps that a technician should perform on a compressor weekly.

_____

_____

_____

13. List the preventive maintenance steps that a technician should perform on a compressor monthly.

_____

_____

_____

INSTRUCTOR COMMENTS:

_____

_____

_____

_____

_____

_____

SPRAY BOOTHS

Name _____   Date _____

Class _____   Instructor _____   Grade _____

This section identifies specific paint department tools and their use and maintenance.

1. In the space provided below, list the personal and environmental safety precautions necessary for this assignment.

   _____

   _____

   _____

2. Why use a spray booth?

   _____

   _____

   _____

3. Describe a crossdraft spray booth.

   _____

   _____

   _____

4. Describe a semi-downdraft spray booth.

   _____

   _____

   _____

5. Describe a downdraft spray booth.

   _____

   _____

   _____

6. Why do most collision repair shops use downdraft spray booths?

   _____

   _____

   _____

7. What is meant by balancing the spray booth and why is it important?

   _____

   _____

   _____

8. What is a prep deck and what is it used for?

   _____

   _____

   _____

9. Describe both paint mixing rooms and their use.

   _____

   _____

   _____

10. What is a spray cycle and why is it important?

_____

_____

_____

11. What is a purge cycle and why is it important?

_____

_____

_____

12. What is a bake or cure cycle and why is it important?

_____

_____

_____

INSTRUCTOR COMMENTS:

_____

_____

_____

_____

_____

_____

BOOTH CLEANING AND MAINTENANCE

Name _____ Date _____

Class _____ Instructor _____ Grade _____

This section identifies specific paint department tools and their use and maintenance.

1. After reading the assignment, in the space provided below, list the personal and environmental safety equipment and precautions needed for this assignment.

   _____

   _____

   _____

2. Why is it important to clean and regularly maintain a spray booth?

   _____

   _____

3. How should the walls of a spray booth be maintained?

   _____

   _____

4. What is the CRI rating for booth lights and why is it important?

   _____

   _____

   _____

5. Why are spray booth lights sealed?

   _____

   _____

6. How should the floors of a spray booth be maintained?

   _____

   _____

7. List the locations and purpose of booth filters.

   _____

   _____

8. How will a technician know when to change booth filters?

   _____

   _____

   _____

9. What does a Velometer test in a spray booth?

   _____

   _____

   _____

10. What is a smoke generator used for when testing a spray booth?

    _____

    _____

    _____

11. How is the air supply tested for debris in a paint booth?

    _____

    _____

    _____

    _____

INSTRUCTOR COMMENTS:

_____

_____

_____

_____

_____

_____

_____

SPRAY BOOTHS

Name _____   Date _____

Class _____   Instructor _____   Grade _____

OBJECTIVES

- Know, understand, and use the safety equipment necessary for the task.
- Demonstrate the ability to operate a spray booth:
  - In spray mode
  - In purge mode
  - In cure mode
- Demonstrate the ability to balance a booth.

NATEF TASK CORRELATION

The written and hands-on activities in this chapter satisfy the NATEF High Priority-Individual and High Priority-Group requirements. This section identifies specific paint department tools and their use and maintenance.

Tools and equipment needed (NATEF tool list)

- Safety glasses
- Ear protection
- Pencil and paper
- Spray booth
- Production sheets for coatings
- Infrared light service manual
- Gloves
- Particle mask
- Assorted hand tools
- Spray booth service manual
- Infrared curing lights

PROCEDURE

1. After reading the work order, gather the safety gear needed to complete the task. In the space provided below, list the personal and environmental safety equipment and precautions needed for this assignment. Have the instructor check and approve your plan before proceeding.

   _____
   _____
   _____

INSTRUCTOR'S APPROVAL _____

2. Booth description:

   _____
   _____

3. Power up the spray booth. List your procedure:

   _____
   _____
   _____

4. Check the production sheet for recommendations for spray temperature. List your findings:

_____

_____

_____

5. Set the spray cycle temperature for the coating being used.
6. Check the production sheet for recommendations for purge temperature and time. Record your findings:

_____

_____

_____

7. Set the correct purge temperature and cycle time.
8. Check the production sheet for recommendations for bake cure temperature and time. Record your findings:

_____

_____

_____

9. Set the correct bake/cure time and temperature for the coating being used.
10. With the infrared light provided, check its service manual and set it up to cure primer filler on a fender. Record your procedure:

_____

_____

_____

INSTRUCTOR COMMENTS:

_____

_____

_____

_____

_____

_____

BOOTH CLEANING AND MAINTENANCE

Name _____   Date _____

Class _____   Instructor _____   Grade _____

OBJECTIVES

- Know, understand, and use the safety equipment necessary for the task.
- Demonstrate the ability to develop and carry out a booth cleaning schedule:
  - Clean the floors.
  - Clean the walls.
  - Maintain the lights in a booth.
  - Change the floor filters.
  - Change the exhaust filters.
  - Change inlet filters.
  - Test the booth for proper operation.

NATEF TASK CORRELATION

The written and hands-on activities in this chapter satisfy the NATEF High Priority-Individual and High Priority-Group requirements. This section identifies specific paint department tools and their use and maintenance.

Tools and equipment needed (NATEF tool list)

- Safety glasses
- Ear protection
- Pencil and paper
- Spray booth
- Velometer
- Particle counter

- Gloves
- Particle mask
- Assorted hand tools
- Booth filters
- Smoke generator
- Thermometer

Booth description:

_____

_____

_____

PROCEDURE

1. After reading the work order, gather the safety gear needed to complete the task. In the space provided below, list the personal and environmental safety equipment and precautions needed for this assignment. Have the instructor check and approve your plan before proceeding.

_____

_____

_____

INSTRUCTOR'S APPROVAL _____

2. Develop a standard operation procedure for regular booth maintenance. List it below:

_____

_____

_____

3. Perform routine booth floor maintenance. List the steps below:

_____

_____

_____

4. Perform routine booth wall maintenance. List the steps below:

_____

_____

_____

5. Perform routine booth lighting maintenance. List the steps below:

_____

_____

_____

6. Perform routine booth floor filter maintenance. List the steps below:

_____

_____

_____

7. Perform routine booth exhaust filter maintenance. List the steps below:

_____

_____

_____

8. Perform routine booth inlet filter maintenance. List the steps below:

_____

_____

_____

9. Develop a standard operation procedure for regular booth operation testing. List it below:

_____

_____

_____

10. Perform routine booth testing. List the steps below:

_____

_____

_____

INSTRUCTOR COMMENTS:

_____

_____

_____

_____

_____

Name _____   Date _____

Class _____   Instructor _____   Grade _____

1. *Technician A* says that a piston compressor is often used in collision repair businesses. *Technician B* says that although a rotary compressor is more costly to purchase, it has a higher duty cycle than a piston compressor. Who is correct?
   A. Technician A only
   B. Technician B only
   C. Both Technicians A and B
   D. Neither Technician A nor B

2. *Technician A* says that an air transformer increases the CFM of compressed air. *Technician B* says that an air transformer or regulator compresses air. Who is correct?
   A. Technician A only
   B. Technician B only
   C. Both Technicians A and B
   D. Neither Technician A nor B

3. *Technician A* says that an extractor cleans dirt out of compressed air. *Technician B* says that an extractor cleans water and oil out of compressed air. Who is correct?
   A. Technician A only
   B. Technician B only
   C. Both Technicians A and B
   D. Neither Technician A nor B

4. *Technician A* says that moisture is not common in a compressed air system. *Technician B* says that moisture can be controlled with a desiccant dryer. Who is correct?
   A. Technician A only
   B. Technician B only
   C. Both Technicians A and B
   D. Neither Technician A nor B

5. *Technician A* says that an aftercooler controls moisture in a compressed air system. *Technician B* says that a dump trap controls moisture in a compressed air system. Who is correct?
   A. Technician A only
   B. Technician B only
   C. Both Technicians A and B
   D. Neither Technician A nor B

6. *Technician A* says that double braded hose is normally used for high-pressure work. *Technician B* says that all hoses are designed to have either air or liquid run through them. Who is correct?
   A. Technician A only
   B. Technician B only
   C. Both Technicians A and B
   D. Neither Technician A nor B

7. *Technician A* says that if the pressure is correct, the length or size of an air hose will be correct. *Technician B* says that a ¼-inch ID air hose is best for painting with an HVLP gun because it is easier to move around in the booth. Who is correct?
   A. Technician A only
   B. Technician B only
   C. Both Technicians A and B
   D. Neither Technician A nor B

8. *Technician A* says that hoses in a booth need to be replaced only if they break. *Technician B* says that both heat and overspray may force the booth hose to be no longer used in a booth. Who is correct?
   A. Technician A only
   B. Technician B only
   C. Both Technicians A and B
   D. Neither Technician A nor B

9. *Technician A* says that regular preventive maintenance is essential to have a reliable long-lasting air compressor. *Technician B* says that the oil level in a compressor should be checked daily. Who is correct?
   A. Technician A only
   B. Technician B only
   C. Both Technicians A and B
   D. Neither Technician A nor B

10. *Technician A* says that pump valves remove debris from a compressor's holding tank. *Technician B* says that if maintenance is performed shortly after the compressor has run, it may have hot parts and caution should be maintained. Who is correct?
    A. Technician A only
    B. Technician B only
    C. Both Technicians A and B
    D. Neither Technician A nor B

11. *Technician A* says that spray booths protect other workers from toxic contamination. *Technician B* says that spray booths provide a controlled environment in which to to spray. Who is correct?
    A. Technician A only
    B. Technician B only
    C. Both Technicians A and B
    D. Neither Technician A nor B

12. *Technician A* says that a semi-downdraft booth has air exhaust filters in the floor. *Technician B* says that a downdraft booth has exhaust filters in the floor. Who is correct?
    A. Technician A only
    B. Technician B only
    C. Both Technicians A and B
    D. Neither Technician A nor B

13. *Technician A* says that balancing a booth to slightly positive airflow will help with contamination. *Technician B* says that balancing a booth to slightly negative airflow will help with contamination. Who is correct?
    A. Technician A only
    B. Technician B only
    C. Both Technicians A and B
    D. Neither Technician A nor B

14. *Technician A* says that when a booth is in spray mode, it recycles the airflow back into the booth. *Technician B* says that when a booth is in bake or cure mode, it recycles air back into the booth. Who is correct?
    A. Technician A only
    B. Technician B only
    C. Both Technicians A and B
    D. Neither Technician A nor B

15. *Technician A* says that baking/curing makes paint catalyze or crosslink. *Technician B* says that baking is what activates paint hardeners. Who is correct?
    A. Technician A only
    B. Technician B only
    C. Both Technicians A and B
    D. Neither Technician A nor B

16. *Technician A* says that a clean booth helps ensure a clean paint job. *Technician B* says that the time spent on maintaining a clean spray area will reduce the time spent on detailing the vehicle later. Who is correct?
    A. Technician A only
    B. Technician B only
    C. Both Technicians A and B
    D. Neither Technician A nor B

17. *Technician A* says that once dust and debris are on the floor, it will not get into the paint. *Technician B* says that the booth should be cleaned before each paint job. Who is correct?
    A. Technician A only
    B. Technician B only
    C. Both Technicians A and B
    D. Neither Technician A nor B

18. *Technician A* says that regular maintenance will make the cleaning of a booth before each job faster. *Technician B* says that if the booth will not balance, the filters may be in need of changing. Who is correct?
    A. Technician A only
    B. Technician B only
    C. Both Technicians A and B
    D. Neither Technician A nor B

19. *Technician A* says that filters should be changed every month. *Technician B* says that filters should be changed checked and changed if necessary when the booths will not balance. Who is correct?
    A. Technician A only
    B. Technician B only
    C. Both Technicians A and B
    D. Neither Technician A nor B

20. *Technician A* says that CRI stands for correct radiant indicator. *Technician B* says that CRI is a measurement of volume of air circulating in a spray booth. Who is correct?
    A. Technician A only
    B. Technician B only
    C. Both Technicians A and B
    D. Neither Technician A nor B

Chapter 23

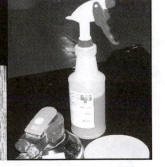

Detailing

■ WORK ASSIGNMENT 23-1

SURFACE INSPECTION

Name _____   Date _____

Class _____   Instructor _____   Grade _____

NATEF IV PAINTING AND REFINISHING, E. 1, 2, 6, 9, 12, 14, 16, 17, 18, 20, 21, 22, 23, 24, 25, 26 27, 28

1. After reading the assignment, in the space provided below, list the personal and environmental safety equipment and precautions needed for this assignment.

   _____
   _____
   _____

In the space provided, describe the repair process necessary to eliminate the paint defects listed below.

2. Inspection: Evaluate the surface for defects or imperfections that will need to be removed and mark with a small piece of tape. Check for:

   Blistering

   Blushing

   Clouding

   Solvent popping

3. Record your findings:

   _____
   _____
   _____

4. Prepare a repair plan.

   _____
   _____
   _____

5. Inspection: Evaluate the surface for defects or imperfections that will need to be removed and mark with a small piece of tape. Check for:

   Contour mapping

   Tape tracking

   Poor adhesion

   Die-back

   Paint cracking

6. Record your findings:

_____

_____

_____

7. Prepare a repair plan.

_____

_____

_____

8. Inspection: Evaluate the surface for defects or imperfections that will need to be removed and mark with a small piece of tape. Check for:
Corrosion

Water spotting

Bird dropping

Airborne finish damage

9. Record your findings:

_____

_____

_____

10. Prepare a repair plan.

_____

_____

_____

11. Inspection: Evaluate the surface for defects or imperfections that will need to be removed and mark with a small piece of tape. Check for:
Chalking

Bleed-through

Pinholes

Buffer marks

Pigment flotation

12. Record your findings:

_____

_____

_____

13. Prepare a repair plan.

_____

_____

_____

INSTRUCTOR COMMENTS:

_____

_____

_____

_____

_____

_____

BUFFING AND POLISHING FINISH TO REMOVE DEFECTS AS REQUIRED

Name _____   Date _____

Class _____   Instructor _____   Grade _____

NATEF IV PAINTING AND REFINISHING, E. 3, 4, 5, 7, 8, 10, 11, 13, 15, 19, 29; F. 2, 3, 4, 5

1. After reading the assignment, in the space provided below, list the personal and environmental safety equipment and precautions needed for this assignment.

_____

_____

2. Inspection: Evaluate the surface for defects or imperfections that will need to be removed and mark with a small piece of tape. Record your findings:

_____

_____

3. Prepare a repair plan; include processes to performed, tools and materials needed, and estimated time for completion. Record your findings:

_____

_____

4. List the tools, equipment, and materials needed for the task. List the needed tools:

_____

_____

5. Measure film thickness prior to detailing. Record your findings:

_____

_____

6. Identify dirt/dust if present and outline the removal process. Record your findings:

_____

_____

7. Identify orange peel if present and outline the removal process. Record your findings:

_____

_____

8. Identify overspray if present and outline the removal process. Record your findings:

_____

_____

9. Identify sags and runs if present and outline the removal process. Record your findings:

   _____

   _____

10. Identify sand scratches if present and outline the removal process. Record your findings:

    _____

    _____

11. Identify color mismatch if present and outline the removal process. Record your findings:

    _____

    _____

12. Identify dry spray if present and outline the removal process. Record your findings:

    _____

    _____

13. Identify fisheye if present and outline the removal process. Record your findings:

    _____

    _____

14. Identify lift if present and outline the removal process. Record your findings:

    _____

    _____

15. Identify any other defects if present and outline the removal process. Record your findings:

    _____

    _____

    _____

INSTRUCTOR COMMENTS:

_____

_____

_____

_____

_____

_____

CLEAN INTERIOR, EXTERIOR, AND GLASS

Name _____   Date _____

Class _____   Instructor _____   Grade _____

NATEF IV, PAINTING AND REFINISHING, F. 3, 4

1. After reading the assignment, in the space provided below, list the personal and environmental safety equipment and precautions needed for this assignment.

   _____

   _____

   _____

On a vehicle, inspect, evaluate, make a work plan, and complete the task needed to clean the interior, exterior, and glass.

2. Inspection: Evaluate the vehicle's interior and exterior, including trim and others for dirt, grime, and polish residue; inspect the vehicle's glass, both interior and exterior. Record your findings:

   _____

   _____

   _____

3. Prepare a repair plan; include processes to performed, tools and materials needed, as well as estimated time for completion. Record your findings:

   _____

   _____

   _____

4. List the tools equipment and materials needed for the task. List the needed tools:

   _____

   _____

   _____

5. Describe the process of cleaning the interior of a vehicle. Record your findings:

   _____

   _____

   _____

6. Describe the process of cleaning the exterior of the vehicle, including the wheels and tires, which should be cleaned at this time as well. Record your findings:

   _____

   _____

   _____

7. Describe the process of glass cleaning, both interior and exterior. Record your findings:

   _____

   _____

   _____

8. Explain why final inspections are important for both the cleaning of the vehicle but also the final inspection for the repairs performed. Record your findings:

_____

_____

_____

INSTRUCTOR COMMENTS:

_____

_____

_____

_____

_____

_____

BUFFING AND POLISHING FINISH TO REMOVE DEFECTS AS REQUIRED

Name _____ Date _____

Class _____ Instructor _____ Grade _____

OBJECTIVES:

- Know, understand, and use the safety equipment necessary for the task.
- Evaluate newly refinished repair.
- Identify defects and imperfections.
- Prepare a repair plan.
- Remove defects and imperfections.
- Complete final inspection.
- Review the work order and vehicle for job completion.
- Send the vehicle to the wash area for final cleaning.

NATEF TASK CORRELATION

The written and hands-on activities in this chapter satisfy the NATEF High Priority-Individual and High Priority-Group requirements for Section IV, Painting and Refinishing, E. 1, 2, 6, 9, 12, 14, 16, 17, 18, 20, 21, 22, 23, 24, 25, 26, 27, 28.

Tools and equipment needed (NATEF tool list)

- Safety glasses
- Ear protection
- Apron
- Wash mitt
- Two wash buckets
- Wax and grease remover
- Tape for marking
- Sanding blocks
- Polishing compounds
- Gloves
- Partial mask
- Pencil and paper
- Automotive soap
- Drying towel or schemes
- Lit magnifying glass
- Sandpaper (P800,1000,1500,2000)
- Buffer
- Polishing bonnets

Vehicle Description

Year_____ Make _____ Model _____

VIN _____ Paint Code _____

Instructions

On a recently completed paint job, you will buff and/or polish it to remove any defect or imperfection that may exist, creating an undetectable repair and preparing it for return to the customer.

PROCEDURE

1. After reading the work order, gather the safety gear needed to complete the task. In the space provided below, list the personal and environmental safety equipment and precautions needed for this assignment. Have the instructor check and approve your plan before proceeding.

_____

_____

_____

INSTRUCTOR'S APPROVAL _____

2. Inspection: Read the work order. Is the work complete?

_____

_____

_____

3. Has the work been done to the shop's standard of quality? If not, should it be sent back for repairs?

_____

_____

_____

4. Inspect the vehicle for defects or imperfections. Mark with tape on the vehicle and list below.

_____

_____

_____

5. Can the defects or imperfections be removed by detailing?

_____

_____

_____

Bleeding

6. Prepare a repair plan; remember, least aggressive first.

Step 1:

_____

_____

_____

Step 2:

_____

_____

_____

Step 3:

_____

_____

_____

Step 4:

_____

_____

_____

Step 5:

_____

_____

_____

7. Remove the defects and imperfections listed in step 2.

8. After removing the noted defects, reexamine the vehicle, list any additional defects, and remove as needed. Record your findings:

_____

_____

_____

9. Have the vehicle inspected by the instructor for final evaluation.

INSTRUCTOR COMMENTS:

_____

_____

_____

_____

_____

_____

CLEAN INTERIOR, EXTERIOR, AND GLASS

Name _____ Date _____

Class _____ Instructor _____ Grade _____

OBJECTIVES

- Know, understand, and use the safety equipment necessary for the task.
- Evaluate the newly polished vehicle.
- Identify special cleaning needs.
- Prepare a repair plan.
- Clean the interior.
- Clean the exterior.
- Clean the glass.
- Complete the final inspection.
- Review the work order and vehicle for job completion.

NATEF TASK CORRELATION

The written and hands-on activities in this chapter satisfy the NATEF High Priority-Individual and High Priority-Group requirements for Section IV, Painting and Refinishing, E. 3, 4, 5, 7, 8, 10, 11, 13, 15, 19, 29; F. 2, 3, 4, 5.

Tools and equipment needed (NATEF tool list)

- Safety glasses
- Ear protection
- Pencil and paper
- Wash mitt
- Two wash buckets
- Wax and grease remover
- Tape for marking
- Detail brushes
- Detail clay lubricant
- Gloves
- Particle mask
- Apron
- Automotive soap
- Drying towel or schemes
- Lit magnifying glass
- Vacuum
- Detail clay

Vehicle Description

Year_____ Make _____ Model _____

VIN _____ Paint Code _____

Instructions

On a recently completed paint job, you will buff and/or polish it to remove any defect or imperfection that may exist, creating an undetectable repair and preparing it for return to the customer.

PROCEDURE

1. After reading the work order, gather the safety gear needed to complete the task. In the space provided below, list the personal and environmental safety equipment and precautions needed for this assignment. Have the instructor check and approve your plan before proceeding.

_____

_____

_____

INSTRUCTOR'S APPROVAL _____

2. Inspect the vehicle for paint defects that can be repaired without refinishing.

3. Has the work been done to the shop's standard of quality? List the defects.

_____

_____

_____

4. Prepare a repair plan; remember, least aggressive first.

   Step 1:

   _____

   _____

   Step 2:

   _____

   _____

   Step 3:

   _____

   _____

   Step 4:

   _____

   _____

   Step 5:

   _____

   _____

INSTRUCTOR COMMENTS:

_____

_____

_____

_____

_____

_____

CLEAN BODY OPENINGS (DOOR JAMBS AND EDGES, ETC.)

Name _____ Date _____

Class _____ Instructor _____ Grade _____

OBJECTIVES

- Know, understand, and use the safety equipment necessary for the task.
- Evaluate the newly polished vehicle.
- Identify special cleaning needs.
- Prepare a repair plan.
- Clean door jambs.
- Clean the trunk opening.
- Clean the gas filler area as needed.
- Clean the engine bay.
- Clean the panel edges as needed.
- Complete the final inspection.
- Review the work order and vehicle for job completion.

NATEF TASK CORRELATION

The written and hands-on activities in this chapter satisfy the NATEF High Priority-Individual and High Priority-Group requirements for Section IV, Painting and Refinishing, F. 3, 4.

Tools and equipment needed (NATEF tool list)

- Safety glasses
- Ear protection
- Apron
- Automotive soap
- Drying towel or schemes
- Lit magnifying glass
- Vacuum
- Detail clay
- Gloves
- Pencil and paper
- Wash mitt
- Two wash buckets
- Wax and grease remover
- Tape for marking
- Detail brushes
- Detail clay lubricant

Vehicle Description

Year_____ Make _____ Model _____

VIN _____ Paint Code _____

Instructions

On a recently completed paint job, you will buff and/or polish it to remove any defect or imperfection that may exist, creating an undetectable repair and preparing it for return to the customer.

PROCEDURE

1. After reading the work order, gather the safety gear needed to complete the task. In the space provided below, list the personal and environmental safety equipment and precautions needed for this assignment. Have the instructor check and approve your plan before proceeding.

_____

_____

_____

INSTRUCTOR'S APPROVAL _____

2. Read the work order. Is it complete?

_____

_____

3. Has the work been done to the shop's standard of quality? If not, should it be sent back for repairs?

_____

_____

_____

4. Prepare a repair plan, remember least aggressive first.

Step 1:

_____

_____

Step 2:

_____

_____

Step 3:

_____

_____

Step 4:

_____

_____

Step 5:

_____

_____

5. Clean the door jambs. Is there a visible line or overspray in the jamb?

_____

_____

_____

6. Clean the trunk opening. Is there a visible line or overspray in the jamb?

_____

_____

7. Clean the gas filler door as needed. Is there a visible line or overspray in the jamb?

_____

_____

_____

8. Clean the engine bay as needed. Is there a visible line or overspray in the jamb?

_____

_____

_____

9. Clean the panel edges as needed. Is there a visible line or overspray in the jamb?

_____

_____

_____

10. Complete the final inspection. Is there any additional work to be completed?

_____

_____

_____

11. Have the vehicle inspected by the instructor for final evaluation.

INSTRUCTOR COMMENTS:

_____

_____

_____

_____

_____

_____

Name _____   Date _____

Class _____   Instructor _____   Grade _____

1. A common paint defect that can be repaired by buffing is:
   A. lifting
   B. couture mapping
   C. orange peel
   D. bull's eyes

2. Overspray on a windshield should be removed with:
   A. 0000 steel wool
   B. thinner
   C. wax and grease remover
   D. detail clay with a lubricant

3. **TRUE** or **FALSE**: Sags and runs can sometimes be removed by detailing.

4. Personal protective devices that should be used when polishing are:
   A. safety glasses
   B. gloves
   C. partial mask
   D. all of the above

5. **TRUE** or **FALSE**: A repair plan helps a technician develop a thorough and efficient process.

6. *Technician A* says that the use of a lighted magnifying glass helps evaluate defects. *Technician B* says that sanding and polishing are required on all finishes. Who is correct?
   A. Technician A only
   B. Technician B only
   C. Both Technicians A and B
   D. Neither Technician A nor B

7. *Technician A* says that all runs must be sanded down and refinished. *Technician B* says that with care a technician can sand and buff out many runs. Who is correct?
   A. Technician A only
   B. Technician B only
   C. Both Technicians A and B
   D. Neither Technician A nor B

8. *Technician A* says that when using a buffer, a partial mask should be worn to keep from inhaling the polishing dust that is spun into the air. *Technician B* says that to help keep clothing clean from splattering compound, an apron should be worn. Who is correct?
   A. Technician A only
   B. Technician B only
   C. Both Technicians A and B
   D. Neither Technician A nor B

9. **TRUE** or **FALSE**: Even though buffers may not seem overly loud, ear protection should be worn.

10. *Technician A* says that if chosen correctly, a single bonnet can be used throughout the polishing steps. *Technician B* says that some bonnets are more aggressive others. Who is correct?
    A. Technician A only
    B. Technician B only
    C. Both Technicians A and B
    D. Neither Technician A nor B

11. *Technician A* says that any liquid soap will adequately clean a vehicle. *Technician B* says that liquid automotive soap should be used to wash a vehicle. Who is correct?
    A. Technician A only
    B. Technician B only
    C. Both Technicians A and B
    D. Neither Technician A nor B

12. **TRUE** or **FALSE**: The two-bucket method of washing uses one bucket of clean, nonsoapy water to rinse out debris from the wash mitt.

13. *Technician A* says that swirl marks are natural and unavoidable. *Technician B* says that swirl marks can be eliminated with "fill and glaze" compounds. Who is correct?
    - **A.** Technician A only
    - **B.** Technician B only
    - **C.** Both Technicians A and B
    - **D.** Neither Technician A nor B

14. In your own words, explain a buffing technique.

    _____

    _____

    _____

15. A vehicle has a film thickness of 3.5 mils before refinishing. After refinishing, it has an average film thickness of 9 mils. The average amount of basecoat is 2 mils. How much of the 9 mils of film thickness can be removed to maintain 2 mils of new clearcoat?

    _____

16. *Technician A* says that detail clay and children's modeling clay are the same. *Technician B* says that detail clay must be used with a lubricant. Who is correct?
    - **A.** Technician A only
    - **B.** Technician B only
    - **C.** Both Technicians A and B
    - **D.** Neither Technician A nor B

17. *Technician A* says that a vehicle should be cleaned from the inside out. *Technician B* says that items that can be taken out to clean should be taken out. Who is correct?
    - **A.** Technician A only
    - **B.** Technician B only
    - **C.** Both Technicians A and B
    - **D.** Neither Technician A nor B

18. *Technician A* says that door jambs may have overspray inside. *Technician B* says that tape over the door gaps helps keep compound out of the door jamb when polishing. Who is correct?
    - **A.** Technician A only
    - **B.** Technician B only
    - **C.** Both Technicians A and B
    - **D.** Neither Technician A nor B

19. *Technician A* says that the engine bay should be checked for compound debris and cleaned before delivery. *Technician B* says that compound debris is often found on the underside of the hood after polishing. Who is correct?
    - **A.** Technician A only
    - **B.** Technician B only
    - **C.** Both Technicians A and B
    - **D.** Neither Technician A nor B

20. *Technician A* says that to remove overspray from plastic headlights, thinner should be used. *Technician B* says that removing overspray from plastic headlights should be done with detail clay. Who is correct?
    - **A.** Technician A only
    - **B.** Technician B only
    - **C.** Both Technicians A and B
    - **D.** Neither Technician A nor B

Chapter 24

Understanding Refinishing

■ WORK ASSIGNMENT 24-1

PAINT CODE RETRIEVAL AND CONVERSION

Name _____ Date _____

Class _____ Instructor _____ Grade _____

NATEF IV PAINTING AND REFINISHING A. 1-6; B. 2

1. After reading the assignment, in the space provided below, list the personal and environmental safety equipment and precautions needed for this assignment.

   _____

   _____

   _____

2. Describe the characteristics of the paint types below.

   Enamel:

   _____

   _____

   _____

   Lacquer:

   _____

   _____

   _____

3. What is the name of the test that can determine a lacquer coating from a nonlacquer one?

   _____

   _____

   _____

4. With the materials provided, perform the test and determine which one is a lacquer finish. Describe your conclusion.

   Sample 1:

   _____

   _____

   _____

   Sample 2:

   _____

   _____

   _____

Sample 3:

_____

_____

Sample 4:

_____

_____

Sample 5:

_____

_____

5. In enamel-based paints, what are the functions of the three components?
Chemically drying rosins:

_____

_____

Pigments:

_____

_____

Solvents:

_____

_____

6. What is the function of a paint hardener?

_____

_____

7. Explain what a thermoplastic coating is.

_____

_____

8. Explain what a thermoset coating is.

_____

_____

_____

INSTRUCTOR COMMENTS:

_____

_____

_____

_____

_____

_____

COMPONENTS AND FUNCTION OF FINISH

Name _____ Date _____

Class _____ Instructor _____ Grade _____

NATEF IV PAINTING AND REFINISHING A. 1-6

1. After reading the assignment, in the space provided below, list the personal and environmental safety equipment and precautions needed for this assignment.

   _____

   _____

   _____

2. Describe the function and safety precautions that should be followed when working with a paint catalyst.

   _____

   _____

   _____

3. How do water-based coatings differ from solvent-based paint?

   _____

   _____

   _____

4. Draw a diagram of the components of solvent-based paint.

5. Draw a diagram of the components of water-based paint.

6. What is another word for rosin and what function does it perform?

   _____

   _____

   _____

7. What function do pigments serve? Explain.

_____

_____

8. What function do solvents serve? Explain.

_____

_____

9. What function do additives serve? Explain.

_____

_____

10. What are the three standard conditions for solvent paint?
    Humidity:

_____

_____

    Temperature:

_____

_____

    Airflow:

_____

_____

11. Explain the 15°F rule.

_____

_____

12. What are value, hue, and chroma?
    Value:

_____

_____

    Hue:

_____

_____

    Chroma:

_____

_____

13. Explain reflection.

_____

_____

_____

14. Explain refraction.

_____

_____

_____

INSTRUCTOR COMMENTS:

_____

_____

_____

_____

_____

_____

■ WORK ASSIGNMENT 24-3

LOCATING OEM PAINT CODE AND PAINT THEORY

Name _____ Date _____

Class _____ Instructor _____ Grade _____

NATEF IV PAINTING AND REFINISHING D. 1, 2

1. After reading the assignment, in the space provided below, list the personal and environmental safety equipment and precautions needed for this assignment.

   _____

   _____

   _____

2. With the aid of the chart provided, explain where to find the OEM code on the vehicle listed.

   Toyota:

   _____

   _____

   Ford:

   _____

   _____

   GM:

   _____

   _____

   BMW:

   _____

   _____

3. How is the OEM code converted to the paint manufacturer's code?

   _____

   _____

4. What is a paint manufacturer's paint formula?

   _____

   _____

5. What is meant by a solid color?

   _____

   _____

6. What is meant by a metallic color?

   _____

   _____

7. What is meant by a prismatic color?

_____

_____

_____

8. What is a refinish system?

_____

_____

_____

9. What is single-stage paint?

_____

_____

_____

10. What is basecoat/clearcoat paint?

_____

_____

_____

11. What is a multistage paint?

_____

_____

_____

INSTRUCTOR COMMENTS:

_____

_____

_____

_____

_____

_____

PAINT CODE RETRIEVAL AND CONVERSION

Name _____ Date _____

Class _____ Instructor _____ Grade _____

OBJECTIVES

- Know, understand, and use the safety equipment necessary for the task.
- Determine and use the personal protective equipment that will be needed for the task.
- Locate the OEM code on the assigned vehicle.
- Convert the OEM code to the paint system code.
- Retrieve the paint formula.
- Prepare the scale for accurate use.

NATEF TASK CORRELATION

The written and hands-on activities in this chapter satisfy the NATEF High Priority-Individual and High Priority-Group requirements for Section IV, Painting and Refinishing, D. 1, 2.

Tools and equipment needed (NATEF tool list)

- Pen and pencil
- Safety glasses
- Respirator
- Gloves
- Ear protection
- MSDS book
- Manufacturer's paint formula retrieval system
- Mixing scale

Vehicle description

Vehicles of differing manufacturers will be provided from which to retrieve OEM paint codes.

PROCEDURE

1. After reading the work order, gather the safety gear needed to complete the task. In the space provided below, list the personal and environmental safety equipment and precautions needed for this assignment. Have the instructor check and approve your plan before proceeding.

_____

_____

_____

INSTRUCTOR'S APPROVAL _____

Retrieve the OEM code from the vehicles provided.

2. Vehicle #1

   Paint code location: _____

   Color description: _____

   OEM code:_____

   Condition of finish:_____

3. Vehicle #2

   Paint code location: _____

   Color description: _____

   OEM code:_____

   Condition of finish:_____

4. Vehicle #3

   Paint code location: _____

   Color description: _____

   OEM code:_____

   Condition of finish:_____

5. Vehicle #4

   Paint code location: _____

   Color description: _____

   OEM code:_____

   Condition of finish:_____

6. Vehicle #5

   Paint code location: _____

   Color description: _____

   OEM code:_____

   Condition of finish:_____

Convert the OEM code to the paint manufacturer's code and locate the formula.

7. Vehicle #1

   Paint manufacturer's code: _____

   Color description: _____

   Formula: _____

8. Vehicle #2

   Paint manufacturer's code: _____

   Color description: _____

   Formula: _____

9. Vehicle #3

   Paint manufacturer's code: _____

   Color description: _____

   Formula: _____

10. Vehicle #4

    Paint manufacturer's code: _____

    Color description: _____

    Formula: _____

11. Vehicle #5

    Paint manufacturer's code: _____

    Color description: _____

    Formula: _____

12. Prepare the scale for measuring. Record your procedure:

    _____

    _____

    _____

INSTRUCTOR COMMENTS:

_____

_____

_____

_____

_____

_____

Name _____   Date _____

Class _____   Instructor _____   Grade _____

1. *Technician A* says that the paint code for vehicles is located on the driver's B pillar. *Technician B* says that the paint code is located in the trunk. Who is correct?
   A. Technician A only
   B. Technician B only
   C. Both Technicians A and B
   D. Neither Technician A nor B

2. *Technician A* says that a painter not only needs to locate the paint code, but he or she needs to check for a "variant" color for that code for a good match. *Technician B* says that when comparing a spray out cart for color match, the technician should use sunlight or a color corrected light. Who is correct?
   A. Technician A only
   B. Technician B only
   C. Both Technicians A and B
   D. Neither Technician A nor B

3. *Technician A* says that when pouring a formula in a professional mixing room, safety equipment is optional. *Technician B* says that fumes in a paint mixing room may be higher than in the spray booth and a respirator is needed. Who is correct?
   A. Technician A only
   B. Technician B only
   C. Both Technicians A and B
   D. Neither Technician A nor B

4. *Technician A* says that keeping the scale clean will help with paint mixing accuracy. *Technician B* says that paint spilled on the scale and not in the cup when mixing a formula will produce a color mismatch. Who is correct?
   A. Technician A only
   B. Technician B only
   C. Both Technicians A and B
   D. Neither Technician A nor B

5. *Technician A* says that a respirator is not needed when spraying waterborne paint. *Technician B* says that protective nitrile gloves should be used when working with paint and solvents. Who is correct?
   A. Technician A only
   B. Technician B only
   C. Both Technicians A and B
   D. Neither Technician A nor B

6. *Technician A* says that the choice of reducers is not that critical so only having one will do. *Technician B* says that the proper choice of reducer helps with finish "flow out" and choosing the correct one is critical. Who is correct?
   A. Technician A only
   B. Technician B only
   C. Both Technicians A and B
   D. Neither Technician A nor B

7. *Technician A* says that the 15°F rule will help a painter calculate pot life. *Technician B* says that for every 15°F temperature increases, cure time is reduced by three times. Who is correct?
   A. Technician A only
   B. Technician B only
   C. Both Technicians A and B
   D. Neither Technician A nor B

8. *Technician A* says that paint is made up of solvent, hardener, and binder. *Technician B* says that paint is made up of solvent, rosin/binder, and pigment. Who is correct?
   A. Technician A only
   B. Technician B only
   C. Both Technicians A and B
   D. Neither Technician A nor B

9. *Technician A* says that if paint comes off on the towel during a solvent test, the finish is a thermoplastic. *Technician B* says that if paint comes off on the towel during a solvent test, the finish is a thermoset. Who is correct?
    A. Technician A only
    B. Technician B only
    C. Both Technicians A and B
    D. Neither Technician A nor B

10. *Technician A* says that e-coat is a thermoset coating. *Technician B* says that e-coat is an excellent corrosion protector. Who is correct?
    A. Technician A only
    B. Technician B only
    C. Both Technicians A and B
    D. Neither Technician A nor B

Chapter 25

Surface Preparation

■ WORK ASSIGNMENT 25-1

INITIAL PREPARATION

Name _____   Date _____

Class _____   Instructor _____   Grade _____

NATEF IV PAINTING AND REFINISHING, B. 1-6, 23

1. After reading the assignment, in the space provided below, list the personal and environmental safety equipment and precautions needed for this assignment.

   _____

   _____

   _____

2. On a vehicle, inspect, evaluate, and make a repair plan to refinish according to its work order. Record your findings:

   _____

   _____

   _____

3. Remove the trim and components necessary for refinishing; bag, tag, and store these components. Record your findings:

   _____

   _____

   _____

Tech Tip

Phone cameras work well for recording placement of trim.

4. Inspect the area to be initially prepared; determine the type of finish and the surface condition. Record your findings:

   _____

   _____

   _____

5. Measure the film thickness to determine if the area should be stripped, or partially stripped, before refinishing. Record your readings:

   1 _____

   2 _____

   3 _____

   4 _____

   5 _____

6. Average film thickness:

   _____

7. Soap and water wash the complete vehicle with automotive soap, and chemically clean the area to be refinished. Record your findings:

   _____

   _____

   _____

8. Remove the paint as determined by the repair plan. Record your findings:

   _____

   _____

   _____

9. Sand the area to be refinished. Record the progression of sandpaper grits:

   _____

   _____

   _____

10. Feather the sanded area, preparing for undercoats. Record your findings:

    _____

    _____

    _____

11. Perform the final inspection and ask the instructor to inspect and evaluate. Record your findings:

    _____

    _____

    _____

INSTRUCTOR COMMENTS:

_____

_____

_____

_____

_____

_____

APPLICATION OF UNDERCOATING

Name _____   Date _____

Class _____   Instructor _____   Grade _____

NATEF IV PAINTING AND REFINISHING, B. 7-14

1. After reading the assignment, in the space provided below, list the personal and environmental safety equipment and precautions needed for this assignment.

   _____

   _____

   _____

2. Inspect the area to have undercoating applied and determine its readiness. Make a repair plan to apply the needed undercoating according to its work order. Record your findings:

   _____

   _____

   _____

3. Mask the area to be undercoated, protecting those areas not to receive undercoating. Record your findings:

   _____

   _____

   _____

4. For this vehicle, apply the appropiate corrosion protection. Record your findings:

   _____

   _____

   _____

5. Determine the correct undercoat for this vehicle (primer filler, primer surfacer, or sealer). Mix the needed amount according to manufacturer recommendations, including its correct color. Record your findings:
   What was used? _____

   What color was used? _____

   What was the amount mixed? _____

6. Apply the correct primer filler, primer surfacer, or sealer. Record your findings:
   What was applied? _____

   How many coats? _____

   What was the outcome?

   _____

   _____

   _____

7. Inspect and apply two-component finishing filler as needed. Record your findings:

   _____

   _____

   _____

8. Block (wet or dry) two-component finishing filler as needed. Record your findings:

_____

_____

_____

DRY POWER BLOCKING

9. Block (wet or dry) the undercoated area. Record your findings:

_____

_____

10. Perform the final inspection and ask the instructor to inspect and evaluate. Record your findings:

_____

_____

_____

INSTRUCTOR COMMENTS:

_____

_____

_____

_____

_____

_____

SEAL FOR PAINT

Name _____   Date _____

Class _____   Instructor _____   Grade _____

NATEF IV PAINTING AND REFINISHING B. 15-21

1. After reading the assignment, in the space provided below, list the personal and environmental safety equipment and precautions needed for this assignment.

   _____

   _____

   _____

2. Inspect the area to have sealed and determine its readiness. Make a repair plan to apply the needed undercoating according to its work order. Record your findings:

   _____

   _____

   _____

3. Inspect the masking for readiness. Record your findings:

   _____

   _____

   _____

4. Clean the area to be sealed with final clean and tack. Record your findings:

   _____

   _____

   _____

5. Select the correct sealer shade and amount, mix to manufacturer's recommendations, and apply to correct coverage. Record your findings:
   Amount mixed: _____

   Mixing ratio: _____

   Shade: _____

   Number of coats: _____

6. De-nib or repair imperfections to prepare for topcoat. Record your findings:

   _____

   _____

   _____

7. Perform the final inspection and ask the instructor to inspect and evaluate. Record your findings:

   _____

   _____

   _____

INSTRUCTOR COMMENTS:

_____
_____
_____
_____
_____
_____

INITIAL PREPARATION

Name _____   Date _____

Class _____   Instructor _____   Grade _____

OBJECTIVES

- Know, understand, and use the safety equipment necessary for the task.
- Inspect, remove, store, and replace the exterior trim and components necessary for proper surface preparation.
- Soap and water wash the entire vehicle, using an appropriate cleaner to remove contaminates.
- Inspect and identify substrate, type of finish, surface condition, and film thickness; develop and document a plan for refinishing using a total product system.
- Remove (as needed) paint finish.
- Dry or wet sand the area to be finished.
- Featheredge the damaged areas to be refinished.

NATEF TASK CORRELATION

The written and hands-on activities in this chapter satisfy the NATEF High Priority-Individual and High Priority-Group requirements for Section IV, Painting and Refinishing, B. 1-6, 23.

Tools and equipment needed (NATEF tool list)

- Safety glasses
- Ear protection
- Pencil and paper
- Apron
- Film thickness gauge
- Automotive soap
- Drying towel or schemes
- Lit magnifying glass
- DA
- Hand sanding pad
- Blowgun
- Gloves
- Particle mask
- Plastic bags
- Assorted hand tools
- Wash mitt
- Two wash buckets
- Wax and grease remover
- Tape for marking
- DA sandpaper P80, 180, 240, and 320
- Wet/dry sandpaper P80, 180, 240, and 320
- Lint-free paper towels

Introduction

Initial preparation of a repaired vehicle restores its corrosion protection and adjusts the surface to the same height as the OEM finish for a seamless and undetectable repair. Being able to prepare a vehicle for refinishing is a critical part of a professional repair.

Vehicle Description

Year_____ Make _____ Model _____

VIN _____ Paint Code _____

PROCEDURE

1. After reading the work order, gather the safety gear needed to complete the task. In the space provided below, list the personal and environmental safety equipment and precautions needed for this assignment. Have the instructor check and approve your plan before proceeding.

_____

_____

_____

INSTRUCTOR'S APPROVAL _____

2. Read the work order. Is it complete?

_____

_____

3. Has the work been done to the shop's standard of quality? If not, should it be sent back for repairs?

_____

_____

4. Inspect, remove, and tag the parts necessary to refinish the vehicle. Store them in a safe place. Number of bags and location of storage:

_____

_____

5. Clean the exterior.
6. Soap and water wash the entire exterior.
7. Chemically clean the area to be refinished.
8. Measure film thickness (measure a minimum of five places and calculate an average).

_____

_____

Average:

_____

9. Identify the substrate, type of finish, and surface condition and prepare a repair plan.

Step 1:

_____

_____

Step 2:

_____

_____

Step 3:

_____

_____

_____

Step 4:

_____

_____

_____

Step 5:

_____

_____

_____

19. Featheredge the area to be prepared.
    Hand sanded

    DA sanded

    Wet

    Dry

    Grit used:_____

11. Have the vehicle inspected by the instructor for final evaluation.

INSTRUCTOR COMMENTS:

_____

_____

_____

_____

_____

_____

APPLICATION OF UNDERCOATING

Name _____ Date _____

Class _____ Instructor _____ Grade _____

OBJECTIVES

- Know, understand, and use the safety equipment necessary for the task.
- Mask for priming.
- Apply corrosion protection.
- Calculate the amount, mix, and reduce the appropriate primer filler or primer surfacer or sealer for the repair.
- Apply primer.
- Block sand (wet or dry) as needed.
- Apply two-component finishing filler as needed.
- Block sand (wet or dry) as needed.
- Clean the area for refinish.

NATEF TASK CORRELATION

The written and hands-on activities in this chapter satisfy the NATEF High Priority-Individual and High Priority-Group requirements for Section IV Painting and Refinishing, B. 7-14.

Tools and equipment needed (NATEF tool list)

- Safety glasses
- Ear protection
- Respirator
- Pencil and paper
- Apron
- Film thickness gauge
- Wax and grease remover
- DA soft pad
- Hand sanding pad
- Sanding paste
- Blowgun
- Masking materials
- 2K primer or sealer
- Cleaning solvent
- Gloves
- Particle mask
- Paint suit
- Plastic bags
- Assorted hand tools
- Wash mitt
- DA
- DA sandpaper P80, 180, 240, 320, 400 and 500
- Wet/dry sandpaper P80, 180, 240, 320, 400, and 500
- Fiber sanding pads (red and gray)
- Lint-free paper towels
- Primer gun
- Reducer

Introduction

Preparing the surface of a vehicle for application of primer filler, primer surface, or sealer as is appropriate for the situation.

Vehicle Description

Year_____ Make _____ Model _____

VIN _____ Paint Code _____

Coating Used _____ Reduction Ratio _____

PROCEDURE

1. After reading the work order, gather the safety gear needed to complete the task. In the space provided below, list the personal and environmental safety equipment and precautions needed for this assignment. Have the instructor check and approve your plan before proceeding.

_____

_____

_____

INSTRUCTOR'S APPROVAL _____

2. Read the work order. Is it complete?

_____

_____

_____

3. Has the work been done to the shop's standard of quality? If not, should it be sent back for repairs?

_____

_____

_____

4. Clean the area and mask area to be primed and make a repair plan.

   Step 1 Corrosion protection primer

   Type: _____

   Amount: _____

   Color: _____

   Reduction: _____

   Number of coats: _____

   Time to topcoat: _____

   Other notes: _____

   _____

   Step 2 Primer filler or surfacer

   Type: _____

   Amount: _____

   Color: _____

   Reduction: _____

   Number of coats: _____

   Time to sanding: _____

   Notes: _____

   _____

   Step 3 Inspection

Step 4 Two-component finish filler

    Mix an appropriate amount of finish filler.

    Smoothly apply the filler.

    Clean the application tools.

    Notes: _____

    _____

Block sanding:

    Apply guide coat.

    Block sand

    Wet

    Dry

    Hand

    Machine

    Abrasives used:

    1. _____

    2. _____

    3. _____

    4. _____

    5. _____

5. Explain why the choices of sanding and the progression of abrasives were used for this project.

    _____

    _____

    _____

6. Clean for topcoating

7. Remove masking.

8. Check for overspray.

9. Remove overspray if necessary.

10. Soap and water wash.

11. Move to the masking area.

12. Have the vehicle inspected by the instructor for final evaluation.

INSTRUCTOR COMMENTS:

_____

_____

_____

_____

_____

_____

SEAL FOR PAINT

Name _____   Date _____

Class _____   Instructor _____   Grade _____

OBJECTIVES

- Know, understand, and use the safety equipment necessary for the task.
- Prepare (according to the work order) the adjacent panels for blending.
- Mask for refinishing.
- Clean the area to be refinished.
- Apply caulking or seam sealer as required.
- Remove dust and other contaminants by tacking.
- Apply stone chip resistant coating as needed.
- Calculate the amount, mix, and reduce the appropriate sealer for the repair.
- Apply sealer.

NATEF TASK CORRELATION

The written and hands-on activities in this chapter satisfy the NATEF High Priority-Individual and High Priority-Group requirements for Section IV, Painting and Refinishing, B. 15-21.

Tools and equipment needed (NATEF tool list)

- Safety glasses
- Ear protection
- Respirator
- Pencil and paper
- DA sandpaper P 400, 500, 600, and 800
- Wet/dry sandpaper P 600, 800, and 320
- Fiber sanding pads (red, gray, and gold)
- Tack cloth
- Lint-free paper towels
- Spray gun
- Reducer
- Gloves
- Particle mask
- Paint suit
- DA with soft pad
- Hand sanding pad
- Sanding past
- Wax and grease remover
- Blowgun
- Masking materials
- 2K sealer
- Cleaning solvent

Introduction

Preparing the surface of a vehicle for application of primer filler, primer surface, or sealer as is appropriate for the situation.

Vehicle Description

Year_____ Make _____ Model _____

VIN _____ Paint Code _____

Coating Used _____ Reduction Ratio _____

PROCEDURE

1. After reading the work order, gather the safety gear needed to complete the task. In the space provided below, list the personal and environmental safety equipment and precautions needed for this assignment. Have the instructor check and approve your plan before proceeding.

    _____

    _____

    _____

INSTRUCTOR'S APPROVAL _____

2. Read the work order. Is it complete?

    _____

    _____

    _____

3. Has the work been done to the shop's standard of quality? If not. should it be sent back for repairs?

    _____

    _____

    _____

4. Prepare the adjacent panels for blending.

    Step 1 Inspect the condition of the finish.
       Scuff the area to be blended.

    Method used: _____

    Abrasives used: _____

5. Explain why the choices of sanding and the progression of abrasives were used for this project.

    _____

    _____

    _____

6. Mask for refinishing. Methods used:

    _____

    _____

    _____

7. Explain why the choices of masking were used for this project.

    _____

    _____

    _____

8. Explain why the choices of sanding and the progression of abrasives were used for this project.

    _____

    _____

    _____

9. Clean for topcoating.
10. Move to the spray area.
11. Apply caulking or seam sealer as required by the work order.

12. Calculate the amount, mix, and reduce the appropriate sealer for the repair.

Type: _____

Amount: _____

Color: _____

Reduction: _____

Number of coats: _____

Flash time: _____

Time to topcoat: _____

Notes: _____

_____

13. Apply sealer.
14. Have the vehicle inspected by the instructor for final evaluation.

INSTRUCTOR COMMENTS:

_____

_____

_____

_____

_____

_____

Name _____ Date _____

Class _____ Instructor _____ Grade _____

1. *Technician A* says that the first step for refinishing preparation is to chemically clean the surface to be refinished. *Technician B* says that the first step for refinishing preparation is soap and water wash. Who is correct?
   A. Technician A only
   B. Technician B only
   C. Both Technicians A and B
   D. Neither Technician A nor B

2. *Technician A* says that guide coat helps a technician identify imperfections. *Technician B* says that guide coat helps level the surface during paint preparation. Who is correct?
   A. Technician A only
   B. Technician B only
   C. Both Technicians A and B
   D. Neither Technician A nor B

3. **True** or **False**: All primers provide corrosion protection.

4. *Technician A* says that adhesion promoter is a coating that helps topcoat stick to OEM paint. *Technician B* says that adhesion promoter is used on certain types of plastic. Who is correct?
   A. Technician A only
   B. Technician B only
   C. Both Technicians A and B
   D. Neither Technician A nor B

5. *Technician A* says that when using a DA sander, the technician should wear gloves. *Technician B* says that a particle mask should be worn when sanding primer filler. Who is correct?
   A. Technician A only
   B. Technician B only
   C. Both Technicians A and B
   D. Neither Technician A nor B

6. *Technician A* says that careful feathering and priming reduces the possibility of the repair showing. *Technician B* says that scuffing and sanding are the same processes. Who is correct?
   A. Technician A only
   B. Technician B only
   C. Both Technicians A and B
   D. Neither Technician A nor B

7. **True** or **False**: Blocking should only be done wet.

8. *Technician A* says that the bare metal should be cleaned with wax and grease remover before applying conversion coating to prevent rust. *Technician B* says that a plastic paint cup or paint cup liner should be used to prevent acid etch primer from harming a steel paint cup. Who is correct?
   A. Technician A only
   B. Technician B only
   C. Both Technicians A and B
   D. Neither Technician A nor B

9. *Technician A* says that each vehicle will have a slightly different refinish plan. *Technician B* says that the first step in cleaning a vehicle for refinishing is to use a wax and grease remover. Who is correct?
   A. Technician A only
   B. Technician B only
   C. Both Technicians A and B
   D. Neither Technician A nor B

10. *Technician A* says that 40-grit sandpaper is the best to use when stripping old finish from aluminum hoods. *Technician B* says that aluminum is used in automobile manufacturing because it does not corrode. Who is correct?
    A. Technician A only
    B. Technician B only
    C. Both Technicians A and B
    D. Neither Technician A nor B

Chapter 26

Masking Materials and Procedures

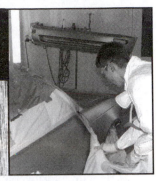

■ WORK ASSIGNMENT 26-1

INSPECT, REMOVE PARTS, AND CLEAN

Name _____ Date _____

Class _____ Instructor _____ Grade _____

NATEF TASK IV B. 1, 2, 3, 8

1. After reading the assignment, in the space provided below, list the personal and environmental safety equipment and precautions needed for this assignment.

   _____

   _____

   _____

2. Describe the importance of reading the work order and making a repair plan. Record your findings:

   _____

   _____

   _____

3. What is the significance of removing parts before masking? Record your findings:

   _____

   _____

   _____

4. If the vehicle has been cleaned multiple times before masking, why should it be washed and chemically cleaned again? Record your findings:

   _____

   _____

   _____

5. Describe a masking repair plan for a complete paint job. Record your findings:

   _____

   _____

   _____

6. Describe a panel mask repair plan. Record your findings:

   _____

   _____

   _____

7. Describe a blend masking repair plan. Record your findings:

_____

_____

_____

8. Describe a masking plan for a tricoat blend. Record your findings:

_____

_____

_____

9. Explain why a vehicle is masked from the inside-out. Record your findings:

_____

_____

_____

INSTRUCTOR COMMENTS:

_____

_____

_____

_____

_____

_____

■ WORK ASSIGNMENT 26-2

MATERIALS AND EQUIPMENT

Name _____   Date _____

Class _____   Instructor _____   Grade _____

NATEF TASK IV B. 1, 2, 3, 8

1. After reading the assignment, in the space provided below, list the personal and environmental safety equipment and precautions needed for this assignment.

   _____

   _____

   _____

2. Describe the importance of using the proper masking paper. Record your findings:

   _____

   _____

3. Why does masking paper come in different sizes? Record your findings:

   _____

   _____

4. What is liquid mask? Record your findings:

   _____

   _____

5. How is plastic drape used, and what is its advantage? Record your findings:

   _____

   _____

6. What are common sizes of masking tape? Record your findings:

   _____

   _____

7. Describe each tape listed below and its use.

   Green:

   _____

   _____

   Blue fine-line:

   _____

   _____

   Beige:

   _____

   _____

8. What is aperture tape used for, and why does it come in different sizes? Record your findings:

_____

_____

_____

9. How does a masking machine work, and how does it speed up the masking process? Record your findings:

_____

_____

_____

INSTRUCTOR COMMENTS:

_____

_____

_____

_____

_____

_____

■ WORK ASSIGNMENT 26-3

TECHNIQUES AND REMOVAL

Name _____ Date _____

Class _____ Instructor _____ Grade _____

NATEF TASKS IV B. 1, 2, 3, 8

1. After reading the assignment, in the space provided below, list the personal and environmental safety equipment and precautions needed for this assignment.

   _____

   _____

   _____

2. Describe the process of masking for protection. Record your findings:

   _____

   _____

   _____

3. Describe how a door jamb would be masked for protection. Record your findings:

   _____

   _____

   _____

4. How would aperture tape be installed into a gas filler door? Record your findings:

   _____

   _____

   _____

5. What special precautions would you take when masking an engine compartment? Record your findings:

   _____

   _____

   _____

6. Describe reverse masking. Record your findings:

   _____

   _____

   _____

7. Describe back masking. Record your findings:

   _____

   _____

   _____

8. When should masking be removed? Record your findings:

   _____

   _____

   _____

9. Why would a technician leave masking material on until after polishing? Record your findings:

   _____

   _____

INSTRUCTOR COMMENTS:

_____

_____

_____

_____

_____

_____

MASKING INSPECTION, PARTS REMOVAL, AND CLEANING

Name _____ Date _____

Class _____ Instructor _____ Grade _____

OBJECTIVES

- Inspect a vehicle to determine how it should be masked for its specific repair.
- Decide which parts should be removed, which parts should be masked, and the method best used to mask them.
- Clean the vehicle for masking.
- Know which masking material to choose for the technique being done.
- Understand what each type of material is used for and how it is best applied.
- Properly handle and maintain the tools used in masking.
- Identify the different techniques used for masking.
- Know how and when to remove masking.

NATEF TASK CORRELATION

The written and hands-on activities in this chapter satisfy the NATEF High Priority-Individual and High Priority-Group requirements for Section IV: Paint and Refinishing, Subsection IV B. 1, 2, 3, 8.

Tools and equipment needed (NATEF tool list)

- Pen and pencil
- Respirator
- Ear protection
- Masking machine with assorted paper
- Liquid mask application gun
- Specialty tape
- Wax and grease remover
- Blowgun
- Safety glasses
- Gloves
- Masking tape, assorted
- Liquid mask
- Plastic drape
- Razor blades
- Paper wipes
- Tack cloths

Instructions

In the lab, on a vehicle that has been prepared for masking, follow the work orders, masking each job as directed.

Year_____ Make _____ Model _____

VIN _____ Paint Code _____

Coating Used _____ Reduction Ratio _____

PROCEDURE

1. After reading the work order, gather the safety gear needed to complete the task. In the space provided below, list the personal and environmental safety equipment and precautions needed for this assignment. Have the instructor check and approve your plan before proceeding.

_____

_____

_____

INSTRUCTOR'S APPROVAL _____

2. Read the work order. Is the work completed?

_____

_____

_____

3. Has the work been done to the shop's standard of quality? If not, should it be sent back for repairs?

_____

_____

_____

4. After reading the work order and inspecting the vehicle, determine if there are any parts remaining on the vehicle that should be removed.

Notes:

_____

_____

_____

5. Bag, tag, and store any removed parts.

Notes:

_____

_____

_____

6. Wash the vehicle with automotive soap and water. Clean the entire vehicle.

Notes:

_____

_____

_____

7. Dry the vehicle and move it to front of the booth for masking.

Notes:

_____

_____

_____

8. Clean the area to be masked with wax and grease remover.

Notes:

_____

_____

_____

INSTRUCTOR COMMENTS:

_____

_____

_____

_____

_____

_____

■ WORK ORDER 26-2

MATERIALS AND EQUIPMENT

Name _____   Date _____

Class _____   Instructor _____   Grade _____

OBJECTIVES

- Inspect a vehicle to determine how it should be masked for its specific repair.
- Decide which parts should be removed, which parts should be masked, and the method best used to mask them.
- Clean the vehicle for masking.
- Know which masking material to choose for the technique being done.
- Understand what each type of material is used for and how it is best applied.
- Properly handle and maintain the tools used in masking.
- Identify the different techniques used for masking.
- Know how and when to remove masking.

NATEF TASK CORRELATION

The written and hands-on activities in this chapter satisfy the NATEF High Priority-Individual and High Priority-Group requirements for Section IV: Paint and Refinishing, Subsection IV B. 1, 2, 3, 8.

Tools and equipment needed (NATEF tool list)

- Pen and pencil
- Respirator
- Ear protection
- Masking machine with assorted paper
- Liquid mask application gun
- Specialty tape
- Wax and grease remover
- Blowgun

- Safety glasses
- Gloves
- Masking tape, assorted
- Liquid mask
- Plastic drape
- Razor blades
- Paper wipes
- Tack cloths

Instructions

In the lab, on the vehicle that has been prepared for masking, follow the work orders, masking each job as directed.

Year_____ Make _____ Model _____

VIN _____ Paint Code _____

Coating Used _____ Reduction Ratio _____

PROCEDURE

1. After reading the work order, gather the safety gear needed to complete the task. In the space provided below, list the personal and environmental safety equipment and precautions needed for this assignment. Have the instructor check and approve your plan before proceeding.

_____

_____

_____

INSTRUCTOR'S APPROVAL _____

2. Read the work order. Is the work completed?

_____

_____

3. Has the work been done to the shop's standard of quality? If not, should it be sent back for repairs?

_____

_____

_____

4. After reading the work order, gather the needed materials to complete the tasks.
   Notes:

_____

_____

5. Mask a vehicle door. Only the exterior is to be painted.
   Instructor comments:

_____

_____

6. Mask a fender for panel painting.
   Instructor comments:

_____

_____

7. Mask a removed mirror.
   Instructor comments:

_____

_____

8. Mask a wheel well.
   Instructor comments:

_____

_____

9. Mask an engine bay.
   Instructor comments:

_____

_____

10. Mask a trunk.
    Instructor comments:

_____

_____

11. Mask a fender for two-toning.

Instructor comments:

_____

_____

_____

INSTRUCTOR COMMENTS:

_____

_____

_____

_____

_____

_____

Name _____ Date _____

Class _____ Instructor _____ Grade _____

1. *Technician A* says that when a new product is introduced into the refinish department, it should be accompanied with the MSDS provided by the manufacturer. *Technician B* says that an MSDS will list the personal protective devices that must be used when working with a product. Who is correct?
   A. Technician A only
   B. Technician B only
   C. Both Technicians A and B
   D. Neither Technician A nor B

2. *Technician A* says that a vehicle that has been repaired in the body shop has been soap and water washed before it was worked on, and therefore it is not necessary to rewash the vehicle when it comes to the refinish department. *Technician B* says that when washing a vehicle, it should be soap and water washed using a two-bucket method. Who is correct?
   A. Technician A only
   B. Technician B only
   C. Both Technicians A and B
   D. Neither Technician A nor B

3. An aftermarket part has arrived in the refinish department. Two technicians have been assured that the black coating is an e-coat primer, but they disagree about how to proceed. *Technician A* says that it is OK, and they should refinish it as they normally do. *Technician B* says that they should perform a wipe-down test to ensure that it is a thermoset product. Who is correct?
   A. Technician A only
   B. Technician B only
   C. Both Technicians A and B
   D. Neither Technician A nor B

4. *Technician A* says that masking is done to protect the parts of the vehicle where new finish is unwanted. *Technician B* says that overspray is not a problem and will not travel far, so covering a vehicle in the back when painting in the front is not necessary. Who is correct?
   A. Technician A only
   B. Technician B only
   C. Both Technicians A and B
   D. Neither Technician A nor B

5. *Technician A* says that a good painter can mask most parts, and not many parts need to be taken off for refinishing. *Technician B* says that the removal of parts will help the refinish technician to refinish the vehicle so the repairs will be undetectable. Who is correct?
   A. Technician A only
   B. Technician B only
   C. Both Technicians A and B
   D. Neither Technician A nor B

6. *Technician A* says that when a vehicle is being masked for panel painting, the finish will be sprayed on the entire panel. *Technician B* says that when masking for blending, the adjacent panels are masked so the color can be blended over the repair and the clear applied to all the exposed panels. Who is correct?
   A. Technician A only
   B. Technician B only
   C. Both Technicians A and B
   D. Neither Technician A nor B

7. *Technician A* says that when preparing a panel for blending of clear, it is masked the same as it would be for blending. *Technician B* says that reverse masking or bridge masking is used to blend clear to eliminate a hard line. Who is correct?
   A. Technician A only
   B. Technician B only
   C. Both Technicians A and B
   D. Neither Technician A nor B

8. *Technician A* says that when masking for priming, overspray is not a problem and 12 inches of paper will be enough to protect the vehicle. *Technician B* says that when masking for priming, it is not necessary to use aperture tape. If primer gets in the cracks, it can be wiped off at any time. Who is correct?
   A. Technician A only
   B. Technician B only
   C. Both Technicians A and B
   D. Neither Technician A nor B

9. *Technician A* says that it matters very little which type of paper is used for masking, so one should use the cheapest. *Technician B* says that paper comes in 36-inch rolls that can be cut down to smaller sizes if needed. Who is correct?
   A. Technician A only
   B. Technician B only
   C. Both Technicians A and B
   D. Neither Technician A nor B

10. *Technician A* says that some papers are resistant to solvents due to their thickness. *Technician B* says that the thinner the paper is, the more conformable it will be. Who is correct?
    A. Technician A only
    B. Technician B only
    C. Both Technicians A and B
    D. Neither Technician A nor B

11. *Technician A* says that most paper machines are very difficult to load, adjust, and work with and, therefore, they are not helpful with masking. *Technician B* says that paper machines come in many different sizes and configurations and greatly speed up the process of masking. Who is correct?
    A. Technician A only
    B. Technician B only
    C. Both Technicians A and B
    D. Neither Technician A nor B

12. *Technician A* says that masking tape backing is made from a paper called crepe. *Technician B* says that tape made from plastic or vinyl helps to eliminate paint from creeping under the tape. Who is correct?
    A. Technician A only
    B. Technician B only
    C. Both Technicians A and B
    D. Neither Technician A nor B

13. *Technician A* says that aperture tapes come in four different sizes to match different size openings. *Technician B* says that trim-masking tape is made to lift flush mount moldings slightly to allow paint to flow underneath the moldings during refinishing. Who is correct?
    A. Technician A only
    B. Technician B only
    C. Both Technicians A and B
    D. Neither Technician A nor B

14. Five identical vehicles come into the shop. The first one requires 625 feet of 12-inch paper to mask it. How many rolls of 12-inch paper will be needed if each roll has 750 feet?

15. Masking tape that measures ¾ inch comes 12 rolls to a sleeve and 4 sleeves to a case. How many rolls are in 12 cases?

Chapter 27

Solvents for Refinishing

■ WORK ASSIGNMENT 27-1

SOLVENT COMPONENTS

Name _____ Date _____

Class _____ Instructor _____ Grade _____

NATEF TASK LIST IV D. 2

1. After reading the assignment, in the space provided below, list the personal and environmental safety equipment and precautions needed for this assignment.

   _____

   _____

   _____

2. Describe how paint solvents are related to both VOCs and HAPs. Record your findings:

   _____

   _____

   _____

3. What purpose do solvents serve in the application of automotive paint? Record your findings:

   _____

   _____

   _____

4. Describe the difference between thinners and reducers. Record your findings:

   _____

   _____

   _____

5. Describe the three phases of solvent evaporation and what purpose they each serve.
   In-flight loss of solvents:

   _____

   _____

   _____

   Leveling solvents:

   _____

   _____

   _____

Tail solvents:

_____

_____

_____

6. What do adhesion promoters do? Record your findings:

_____

_____

_____

7. How does a wax and grease remover work? Record your findings:

_____

_____

_____

8. What does a hardener do? Record your findings:

_____

_____

_____

9. Explain the difference between a thermoset and a thermoplastic paint. Record your findings:

_____

_____

_____

10. What is the 15-degree rule and how does it work? Record your findings:

_____

_____

_____

INSTRUCTOR COMMENTS:

_____

_____

_____

_____

_____

_____

REDUCTION AND MIXING

Name _____   Date _____

Class _____   Instructor _____   Grade _____

NATEF TASK LIST IV D. 2

1. After reading the assignment, in the space provided below, list the personal and environmental safety equipment and precautions needed for this assignment.

_____

_____

_____

2. Describe how each of the conditions below affects a painter's choice of reducers.

    Temperature:

    _____

    _____

    _____

    Humidity:

    _____

    _____

    _____

    Airflow:

    _____

    _____

    _____

    Time:

    _____

    _____

    _____

    Film thickness:

    _____

    _____

    _____

3. Describe how each of the following products works.

    Color blender:

    _____

    _____

    _____

    Accelerator:

    _____

    _____

    _____

Retarder:

_____

_____

_____

4. What is a mixing ratio? Record your findings:

_____

_____

_____

5. How does weight reduction compare to part reduction? Record your findings:

_____

_____

_____

6. What importance do stirring and agitating have? Record your findings:

_____

_____

_____

INSTRUCTOR COMMENTS:

_____

_____

_____

_____

_____

_____

REDUCTION AND MIXING, PART 2

Name _____ Date _____

Class _____ Instructor _____ Grade _____

NATEF TASK LIST IV D. 2

1. After reading the assignment, in the space provided below, list the personal and environmental safety equipment and precautions needed for this assignment.

   _____

   _____

   _____

2. What effect do solvents have on the environment? Record your findings:

   _____

   _____

   _____

3. What are VOCs? Record your findings:

   _____

   _____

   _____

4. What are HAPs? Record your findings:

   _____

   _____

   _____

5. What is the advantage of recycling solvents? Record your findings:

   _____

   _____

   _____

INSTRUCTOR COMMENTS:

_____

_____

_____

_____

_____

_____

■ WORK ORDER 27-1

SOLVENT COMPONENTS

Name _____ Date _____

Class _____ Instructor _____ Grade _____

OBJECTIVES

- Understand how solvents are used in the refinish industry.
- Control the variables that affect the quality and speed of refinishing.
- Choose the proper solvents to use when changes occur in factors such as temperature, humidity, and air movement.
- Understand why certain spray techniques, such as keeping the proper distance from the object being sprayed, are important.
- Understand why different blending techniques work better with certain colors.

NATEF TASK CORRELATION

The written and hands-on activities in this chapter satisfy the NATEF High Priority-Individual and High Priority-Group requirements for Section IV: Paint and Refinishing, Subsection D, Task 2.

Tools and equipment needed (NATEF tool list)

- Pen and paper
- Respirator
- Ear protection
- MSDS
- Measuring cups
- Paint retrieval computer
- Paper towels

- Safety glasses
- Gloves
- Paint suit
- Technical data sheets for paint products
- Scale
- Assorted reducers

Instructions

The lab has been prepared for the mixing and solvent exercises for this section. Read the task, proceed to the workstation, and complete the task as directed.

Vehicle Description

Year_____ Make _____ Model _____

VIN _____ Paint Code _____

Coating Used _____ Reduction Ratio _____

PROCEDURE

1. After reading the work order, gather the safety gear needed to complete the task. In the space provided below, list the personal and environmental safety equipment and precautions needed for this assignment. Have the instructor check and approve your plan before proceeding.

_____

_____

_____

INSTRUCTOR'S APPROVAL _____

2. You are painting in 80°F temperature, 95% humidity, and in a downdraft booth that moves 100 CFM of air past the vehicle you are painting. Which solvent would you choose: a slow, a medium, or a fast solvent?

   _____

3. How does humidity affect the evaporation? Record your findings:

   _____

   _____

   _____

4. You are clear blending a vehicle. How would you use color blender? Record your findings:

   _____

   _____

   _____

5. Under what conditions would you use an accelerator or a retarder? Record your findings:

   _____

   _____

   _____

6. In a mixing cup, mix paint hardener and solvent in a 4-2-1 ratio.

   Instructor comments:

   _____

   _____

   _____

7. Using a scale, mix paint using 350 grams of paint, 100 grams of hardener, and 85 grams of slow solvent.

   Instructor comments:

   _____

   _____

   _____

8. Using a shaker, agitate a reduced paint for 2 minutes.

INSTRUCTOR COMMENTS:

_____

_____

_____

_____

_____

_____

SOLVENT COMPONENTS

Name _____   Date _____

Class _____   Instructor _____   Grade _____

OBJECTIVES

- Understand how solvents are used in the refinish industry.
- Control the variables that affect the quality and speed of refinishing.
- Choose the proper solvents to use when changes occur in factors such as temperature, humidity, and air movement.
- Understand why certain spray techniques, such as keeping the proper distance from the object being sprayed, are important.
- Understand why different blending techniques work better with certain colors.

NATEF TASK CORRELATION

The written and hands-on activities in this chapter satisfy the NATEF High Priority-Individual and High Priority-Group requirements for Section IV: Paint and Refinishing, Subsection D, Task 2.

Tools and equipment needed (NATEF tool list)

- Pen and paper
- Respirator
- Paint suit
- MSDS
- Measuring cups
- Paint retrieval computer
- Paper towels
- Safety glasses
- Gloves
- Ear protection
- Technical data sheets for paint products
- Scale
- Assorted reducers

Instructions

The lab has been prepared for the mixing and solvent exercises for this section. Read the task, proceed to the workstation, and complete the task as directed.

Vehicle Description

Year_____ Make _____ Model _____

VIN _____ Paint Code _____

Coating Used _____ Reduction Ratio _____

PROCEDURE

1. After reading the work order, gather the safety gear needed to complete the task. In the space provided below, list the personal and environmental safety equipment and precautions needed for this assignment. Have the instructor check and approve your plan before proceeding.

_____

_____

_____

INSTRUCTOR'S APPROVAL _____

2. Identify all the locations in the lab where solvents potentially can escape into the atmosphere. Record your findings:

_____

_____

_____

3. Demonstrate the proper operation of a paint booth to reduce solvent evaporation in the shop. Instructor comments:

_____

_____

_____

4. Demonstrate the proper operation of a prep station to reduce solvent evaporation in the shop. Instructor comments:

_____

_____

_____

5. How does a prep station reduce HAPs from getting into the environment? Record your findings:

_____

_____

_____

6. Demonstrate the correct operation of a gun washing machine. Instructor comments:

_____

_____

_____

7. Demonstrate the correct operation of a recycle machine. Instructor comments:

_____

_____

_____

8. Demonstrate how to correctly hand wash a paint gun. Instructor comments:

_____

_____

_____

INSTRUCTOR COMMENTS:

_____

_____

_____

_____

_____

_____

Name _____ Date _____

Class _____ Instructor _____ Grade _____

1. *Technician A* says that the personal safety equipment used when handling solvents is the same as when using engine coolant. *Technician B* says that to find the personal safety equipment needed to handle solvents, a technician should read, understand, and follow the MSDS instructions. Who is correct?
   A. Technician A only
   B. Technician B only
   C. Both Technicians A and B
   D. Neither Technician A nor B

2. *Technician A* says that the first cleaning done to a vehicle before working on it is a chemical cleaning with a wax and grease remover. *Technician B* says that a solvent is a general word for many substances that dissolve and carry another chemical. Who is correct?
   A. Technician A only
   B. Technician B only
   C. Both Technicians A and B
   D. Neither Technician A nor B

3. *Technician A* says that the term "volatile" when talking about VOCs refers to chemicals that evaporate easily. *Technician B* says that organic when talking about VOCs refers to chemicals that contain carbon. Who is correct?
   A. Technician A only
   B. Technician B only
   C. Both Technicians A and B
   D. Neither Technician A nor B

4. *Technician A* says that many of the products that paint technicians use contain solvents. *Technician B* says that paints coming from the manufacturer do not have solvents in them, so a paint technician must add them so the coating can pass through the spray gun properly. Who is correct?
   A. Technician A only
   B. Technician B only
   C. Both Technicians A and B
   D. Neither Technician A nor B

5. *Technician A* says that lacquer paints are thinned. *Technician B* says that enamel and urethane paints are reduced. Who is correct?
   A. Technician A only
   B. Technician B only
   C. Both Technicians A and B
   D. Neither Technician A nor B

6. *Technician A* says that leveling solvents evaporate from paint last, sometimes as long as 90 days after spraying. *Technician B* says that the first solvents to evaporate are leveling solvents. Who is correct?
   A. Technician A only
   B. Technician B only
   C. Both Technicians A and B
   D. Neither Technician A nor B

7. *Technician A* says that the surface tension of cured paint is what makes it scratch-resistant. *Technician B* says that surface tension only exists on liquids. Who is correct?
   A. Technician A only
   B. Technician B only
   C. Both Technicians A and B
   D. Neither Technician A nor B

8. *Technician A* says that before blending new paint onto OEM paint, an adhesion promoter should be applied so the old finish will soften and let the new finish blend into it. *Technician B* says that an adhesion promoter is used to help with the adhesion of a coating on polyolefin plastics. Who is correct?
   A. Technician A only
   B. Technician B only
   C. Both Technicians A and B
   D. Neither Technician A nor B

9. *Technician A* says that wax and grease remover is a chemical cleaner. *Technician B* says that wax and grease remover is a solvent. Who is correct?
   A. Technician A only
   B. Technician B only
   C. Both Technicians A and B
   D. Neither Technician A nor B

10. *Technician A* says that lacquer paint is a two-part paint and is sometimes called 2K paint. *Technician B* says that isocyanides are chemicals used to reduce paints. Who is correct?
    A. Technician A only
    B. Technician B only
    C. Both Technicians A and B
    D. Neither Technician A nor B

11. *Technician A* says that when paint reaches its crosslinked and thermoset stage, it will no longer reflow when solvents are applied to it. *Technician B* says that a thermoplastic paint will soften when solvents are applied to it. Who is correct?
    A. Technician A only
    B. Technician B only
    C. Both Technicians A and B
    D. Neither Technician A nor B

12. *Technician A* says that the four components found in paint shipped from the manufacturer are rosins, pigments, solvents, and additives. *Technician B* says that the way to measure the thickness of a coating is with a viscosity cup. Who is correct?
    A. Technician A only
    B. Technician B only
    C. Both Technicians A and B
    D. Neither Technician A nor B

13. 1525 ml of paint must be reduced at a 125% ratio. How much reducer should be added?

14. If a paint department uses ½ pint of wax and grease remover a day, how many gallons of wax and grease remover would that business need each year?

15. 25 gallons of paint are being reduced at a 4-1-1 ratio. How many quarts of reducer must be added to the paint?

Chapter 28

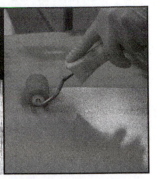

Application of Undercoats

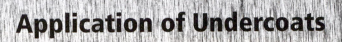

■ WORK ASSIGNMENT 28-1

TOOLS, EQUIPMENT, AND TYPES OF UNDERCOATS

Name _____   Date _____

Class _____   Instructor _____   Grade _____

NATEF TASKS A. 1-6; B. 1-18, 20-23; C. 1-3

1. After reading the assignment, in the space provided below, list the personal and environmental safety equipment and precautions needed for this assignment.

   _____

   _____

2. Describe how a primer gun is set up differently from a basecoat gun or clearcoat gun. Record your findings:

   _____

   _____

   _____

3. Explain the different ways that undercoats can be applied and the advantages and disadvantages of each method.

   Spray:

   _____

   _____

   _____

   Roller:

   _____

   _____

   Brush:

   _____

   _____

   _____

4. Describe how masking for undercoats is different than for topcoats. Record your findings:

   _____

   _____

   _____

5. Describe the different purposes for the undercoats named.
   Metal cleaner/conversion coating:

   _____

   _____

   _____

   Epoxy primers:

   _____

   _____

   _____

   Etching/wash primers:

   _____

   _____

   _____

   Primer (filler):

   _____

   _____

   _____

   Primer-surfacer:

   _____

   _____

   _____

   Sealer:

   _____

   _____

   _____

   Adhesion-promoting primers:

   _____

   _____

   _____

INSTRUCTOR COMMENTS:

_____

_____

_____

_____

_____

_____

■ WORK ASSIGNMENT 28-2

SURFACE PREPARATION AND APPLICATION

Name _____   Date _____

Class _____   Instructor _____   Grade _____

NATEF TASKS A. 1-6; B. 1-18, 20-23; C. 1-3

1. After reading the assignment, in the space provided below, list the personal and environmental safety equipment and precautions needed for this assignment.

   _____
   _____
   _____

2. Describe the steps for using metal conditioner/conversion coating. Record your findings:

   _____
   _____
   _____

3. How does the use of metal conditioner/conversion coating differ when it is applied to aluminum and when it is applied to steel? Record your findings:

   _____
   _____
   _____

4. What is epoxy primer used for? Record your findings:

   _____
   _____
   _____

5. What is primer-surfacer used for? Record your findings:

   _____
   _____
   _____

6. How do primer filler and primer surfacer differ? Record your findings:

   _____
   _____
   _____

7. What is a primer sealer and when is it used? Record your findings:

   _____
   _____
   _____

8. Do you always need to apply sealer before applying a topcoat? Explain. Record your findings:

   _____
   _____
   _____

9. What is an adhesion promoter used for? Record your findings:

   _____
   _____

10. How does masking for undercoats differ from masking for topcoats? Record your findings:

_____

_____

_____

11. How does the application of undercoats differ from the application of topcoats? Record your findings:

_____

_____

_____

INSTRUCTOR COMMENTS:

_____

_____

_____

_____

_____

_____

TOOLS, EQUIPMENT, AND TYPES OF UNDERCOATS

Name _____ Date _____

Class _____ Instructor _____ Grade _____

OBJECTIVES

- Be able to use and apply undercoats safely.
- Understand the different types of undercoating and their use.
- Understand the different types of application of undercoats.
- Be familiar with the tools needed to apply undercoats.
- Be familiar with the differing surface preparations needed for undercoat applications.
- Know when to use the different types of undercoats.

NATEF TASK CORRELATION

The written and hands-on activities in this chapter satisfy the NATEF High Priority-Individual and High Priority-Group requirements for Section IV: Paint and Refinishing, Subsections A. 1-6; B. 1-18, 20-23; C. 1-3.

Tools and equipment needed (NATEF tool list)

- Pen and pencil
- Respirator
- Paint suit
- Masking tape, assorted
- Plastic drape
- Razor blades
- Paper wipes
- Tack cloths
- Role application tools
- Roller pads
- Foam brush
- Sealer gun
- Safety glasses
- Gloves
- Ear protection
- Masking machine with assorted paper
- Specialty tape
- Wax and grease remover
- Blowgun
- Different types of undercoats
- Roller arm
- Roller pan or cup
- Primer gun

Instructions

In the lab, vehicles have been prepared for undercoats. Follow the work orders for each job and proceed as directed.

Vehicle Description

Year_____ Make _____ Model _____

VIN _____ Paint Code _____

Coating Used _____ Reduction Ratio _____

PROCEDURE

1. After reading the work order, gather the safety gear needed to complete the task. In the space provided below, list the personal and environmental safety equipment and precautions needed for this assignment. Have the instructor check and approve your plan before proceeding.

   _____

   _____

   _____

INSTRUCTOR'S APPROVAL _____

2. Gather and list the tools and equipment needed to roll prime, using primer surfacer.
   Notes:

   _____

   _____

   _____

3. Gather and list the tools and equipment to apply epoxy primer, using a foam brush.
   Notes:

   _____

   _____

   _____

4. Gather and list the tools and equipment needed to apply sealer, using a gun.
   Notes:

   _____

   _____

   _____

5. How does a sealer gun differ from a primer gun?
   Notes:

   _____

   _____

   _____

6. Find the technical data sheet for your manufacturer's epoxy primer and list the mixing ratio.
   Notes:

   _____

   _____

   _____

7. Find the technical data sheet for your manufacturer's primer surfacer and list the mixing ratio.
   Notes:

   _____

   _____

   _____

8. Find the technical data sheet for your manufacturer's sealer and list the mixing ratio.
   Notes:

   _____

   _____

   _____

9. Find the technical data sheet for your manufacturer's etch primer and list the mixing ratio.

Notes:

_____

_____

_____

INSTRUCTOR COMMENTS:

_____

_____

_____

_____

_____

_____

SURFACE PREPARATION AND APPLICATION

Name _____   Date _____

Class _____   Instructor _____   Grade _____

OBJECTIVES

- Be able to use and apply undercoats safely.
- Understand the different types of undercoating and their use.
- Understand the different types of application of undercoats.
- Be familiar with the tools needed to apply undercoats.
- Be familiar with the differing surface preparations needed for undercoat applications.
- Know when to use the different types of undercoats.

NATEF TASK CORRELATION

The written and hands-on activities in this chapter satisfy the NATEF High Priority-Individual and High Priority-Group requirements for Section IV: Paint and Refinishing, Subsections A. 1-6; B. 1-18, 20-23; C. 1-3.

Tools and equipment needed (NATEF tool list)

- Pen and pencil
- Respirator
- Paint suit
- Masking tape, assorted
- Plastic drape
- Razor blades
- Paper wipes
- Tack cloths
- Role application tools
- Roller pads
- Foam brush
- Sealer gun
- Safety glasses
- Gloves
- Ear protection
- Masking machine with assorted paper
- Specialty tape
- Wax and grease remover
- Blowgun
- Different types of undercoats
- Roller arm
- Roller pan or cup
- Primer gun

Instructions

In the lab, vehicles have been prepared for undercoats. Follow the work orders for each job and proceed as directed.

Vehicle Description

Year_____ Make _____ Model _____

VIN _____ Paint Code _____

Coating Used _____ Reduction Ratio _____

PROCEDURE

1. After reading the work order, gather the safety gear needed to complete the task. In the space provided below, list the personal and environmental safety equipment and precautions needed for this assignment. Have the instructor check and approve your plan before proceeding.

_____

_____

_____

INSTRUCTOR'S APPROVAL _____

2. Gather the tools needed for application of sprayed primer. Consult the technical data sheet (TDS) and mix primer surfacer, using the proper tint or value for the topcoat.

   Notes:

   _____

   _____

3. Gather the tools needed for application of sprayed primer. Consult the technical data sheet (TDS) and mix primer filler, using the proper tint or value for the topcoat.

   Notes:

   _____

   _____

4. Gather the tools needed for application of sprayed primer. Consult the technical data sheet (TDS) and mix etch primer, using the proper tint or value for the topcoat.

   Notes:

   _____

   _____

5. Gather the tools needed for application of sprayed primer. Consult the technical data sheet (TDS) and mix epoxy primer, using the proper tint or value for the topcoat.

   Notes:

   _____

   _____

6. Gather the tools needed for application of sprayed primer. Consult the technical data sheet (TDS) and mix sealer, using the proper tint or value for the topcoat.

   Notes:

   _____

   _____

7. Gather the tools needed for application of sprayed primer. Consult the technical data sheet (TDS) and mix direct-to-metal primer surfacer, using the proper tint or value for the topcoat.

   Notes:

   _____

   _____

   _____

INSTRUCTOR COMMENTS:

_____

_____

_____

_____

_____

_____

Name _____ **Date** _____

Class _____ **Instructor** _____ **Grade** _____

1. *Technician A* says that the national rule was enacted in 1998. *Technician B* says that 2K paint products require a hardener. Who is correct?
   A. Technician A only
   B. Technician B only
   C. Both Technicians A and B
   D. Neither Technician A nor B

2. *Technician A* says that metal cleaning and conversion coating are no longer required. *Technician B* says that some paint manufacturers may recommend metal cleaning and conversion coating before using their epoxy primers. Who is correct?
   A. Technician A only
   B. Technician B only
   C. Both Technicians A and B
   D. Neither Technician A nor B

3. *Technician A* says that epoxy paint is a primer used as a sealer before spraying topcoat. *Technician B* says that epoxy paint is a primer used to fill minor imperfections. Who is correct?
   A. Technician A only
   B. Technician B only
   C. Both Technicians A and B
   D. Neither Technician A nor B

4. *Technician A* says that some paints may have an "induction period," which is the length of time that the coating must rest after adding the catalyst before spraying. *Technician B* says that primer surfacer does not require sanding before topcoating. Who is correct?
   A. Technician A only
   B. Technician B only
   C. Both Technicians A and B
   D. Neither Technician A nor B

5. *Technician A* says that VOC stands for "volatile organic components." *Technician B* says that an acid etch primer must be sanded before topcoating. Who is correct?
   A. Technician A only
   B. Technician B only
   C. Both Technicians A and B
   D. Neither Technician A nor B

6. *Technician A* says that wash primers contain acid. *Technician B* says that primer surfacer must be sanded before topcoating. Who is correct?
   A. Technician A only
   B. Technician B only
   C. Both Technicians A and B
   D. Neither Technician A nor B

7. *Technician A* says that paint thickness is measured in mils. *Technician B* says that paint booths protect workers who are outside the booth from hazardous overspray. Who is correct?
   A. Technician A only
   B. Technician B only
   C. Both Technicians A and B
   D. Neither Technician A nor B

8. *Technician A* says that roll primer reduces the amount of VOCs vented into the atmosphere. *Technician B* says that roll primer has 75% transfer efficiency. Who is correct?
   A. Technician A only
   B. Technician B only
   C. Both Technicians A and B
   D. Neither Technician A nor B

9. *Technician A* says that adhesion promoters are primers used on polyolefin plastics. *Technician B* says that adhesion promoters are primers used to promote adhesion to OEM paints. Who is correct?
   A. Technician A only
   B. Technician B only
   C. Both Technicians A and B
   D. Neither Technician A nor B

10. *Technician A* says that blending solvents help the new paint melt into the OEM finish. *Technician B* says that blending solvents help with metallic orientation. Who is correct?
    A. Technician A only
    B. Technician B only
    C. Both Technicians A and B
    D. Neither Technician A nor B

11. *Technician A* says that the first step in surface preparation is soap and water washing. *Technician B* says that applying wax and grease remover is the first step in surface preparation. Who is correct?
    A. Technician A only
    B. Technician B only
    C. Both Technicians A and B
    D. Neither Technician A nor B

12. *Technician A* says that aluminum should never be sanded with sandpaper coarser than P180 grit. *Technician B* says that sanded aluminum should be coated within ½ hour of sanding, or the surface will need to be resanded. Who is correct?
    A. Technician A only
    B. Technician B only
    C. Both Technicians A and B
    D. Neither Technician A nor B

13. *Technician A* says that primer sealers provide adhesion to the substrate they are applied to and to the topcoat sprayed on them. *Technician B* says that sealers act as a barrier coat between the old finish under them and the new topcoat being applied to them. Who is correct?
    A. Technician A only
    B. Technician B only
    C. Both Technicians A and B
    D. Neither Technician A nor B

14. *Technician A* says that some primers are tintable. *Technician B* says that some paint manufacturers provide primers that come in different colors, which can be mixed to create an undercoat with the best value for topcoating. Who is correct?
    A. Technician A only
    B. Technician B only
    C. Both Technicians A and B
    D. Neither Technician A nor B

15. *Technician A* says that multipurpose undercoats have been developed to perform like a sealer, a primer filler, or a corrosion protector. *Technician B* says that multipurpose undercoats are primer fillers only. Who is correct?
    A. Technician A only
    B. Technician B only
    C. Both Technicians A and B
    D. Neither Technician A nor B

Chapter 29

Advanced Refinishing Procedures

■ WORK ASSIGNMENT 29-1

APPLICATION OF BASE COAT AND BLENDING

Name _____ Date _____

Class _____ Instructor _____ Grade _____

NATEF TASKS A. 1-6; C. 1-3; D. 1-10

1. After reading the assignment, in the space provided below, list the personal and environmental safety equipment and precautions needed for this assignment.

   _____

   _____

   _____

2. When may brush application of finish be the best method for applying undercoats? Record your findings:

   _____

   _____

   _____

3. When may roll application of a finish be the best method for applying undercoats? Record your findings:

   _____

   _____

   _____

4. How does engine bay application of color coat and clearcoat differ from surface application? Record your findings:

   _____

   _____

   _____

5. Explain the procedure for edging in parts. Record your findings:

   _____

   _____

   _____

6. What is a wet-on-wet sealer, and how is it applied? Record your findings:

   _____

   _____

   _____

7. Explain how to apply a standard blend. Record your findings:

_____

_____

_____

8. Explain how to apply a reverse blend. Record your findings:

_____

_____

_____

9. Explain how to apply a wet-bed blend. Record your findings:

_____

_____

_____

10. What is clear blending? Record your findings:

_____

_____

_____

INSTRUCTOR COMMENTS:

_____

_____

_____

_____

_____

_____

OVERALL, CLEAR BLENDING, AND CLEAR APPLICATION

Name _____   Date _____

Class _____   Instructor _____   Grade _____

NATEF TASKS A. 1-6; C. 1-3; D. 1-10

1. After reading the assignment, in the space provided below, list the personal and environmental safety equipment and precautions needed for this assignment.

   _____

   _____

   _____

2. Can single stage be blended? If so, how? Record your findings:

   _____

   _____

3. Explain how multistage blending differs from basecoat/clearcoat blending. Record your findings:

   _____

   _____

4. Explain the procedure for painting an overall refinish. Record your findings:

   _____

   _____

5. Following the application of color coat, how is clear applied to the panel? Record your findings:

   _____

   _____

6. Is clearcoat always applied in two full coats? If not, explain. Record your findings:

   _____

   _____

   _____

INSTRUCTOR COMMENTS:

_____

_____

_____

_____

_____

■ WORK ASSIGNMENT 29-3

WATERBORNE

Name _____   Date _____

Class _____   Instructor _____   Grade _____

NATEF TASKS A. 1-6; C. 1-3; D. 1-10

1. After reading the assignment, in the space provided below, list the personal and environmental safety equipment and precautions needed for this assignment.

   _____

   _____

2. How does paint storage of waterborne coatings differ from storage of solvent paint? Record your findings:

   _____

   _____

3. How may air needs for waterborne application differ from those of solvent paint? Record your findings:

   _____

   _____

Find the TDS for your paint manufacturer and answer the following questions:

4. How is the surface prepared for waterborne?

   _____

   _____

5. Which undercoat is used, and how is it reduced?

   _____

   _____

6. Is the masking different than with solvent paint?

   _____

   _____

7. Are there any special considerations when using undercoat that differ from solvent application?

   _____

   _____

8. What is the recommended paint gun setup for waterborne? Record your findings:

   _____

   _____

   _____

9. Explain the procedure for waterborne application. Record your findings:

_____

_____

_____

10. What is a control coat, and how is it done? Record your findings:

_____

_____

_____

11. How does flash time for waterborne coatings differ from solvent coatings? Record your findings:

_____

_____

_____

12. Can waterborne be blended? If so, how should it be done? Record your findings:

_____

_____

_____

13. What considerations must be observed when blending multistage paint with waterborne? Record your findings:

_____

_____

_____

14. How is a paint gun with waterborne paint cleaned? Record your findings:

_____

_____

_____

15. How does waterborne waste disposal differ from waste disposal of solvent paint? Record your findings:

_____

_____

_____

16. Explain clear application following waterborne application. Record your findings:

_____

_____

_____

INSTRUCTOR COMMENTS:

_____

_____

_____

_____

_____

_____

APPLICATION OF BASE COAT AND BLENDING

Name _____   Date _____

Class _____   Instructor _____   Grade _____

OBJECTIVES

- Calculate the amount of coating needed for specific applications.
- Understand and be able to paint at high transfer efficiency level.
- Understand and perform techniques of application that will reduce time and material.
- Understand and be able to perform blending techniques such as:
  - Standard blending
  - Reverse blending
  - Wet-bedding blending
  - Zone blending
  - Single-stage blending
- Repair multistage paints.
- Understand waterborne application.

NATEF TASK CORRELATION

The written and hands-on activities in this chapter satisfy the NATEF High Priority-Individual and High Priority-Group requirements for Section IV: Paint and Refinishing, Subsections A. 1-6; C. 1-3; D. 1-10.

Tools and equipment needed (NATEF tool list)

- Pen and pencil
- Respirator
- Paint suit
- Coating technical data sheets (TDS)
- Waterborne coatings
- Wax and grease remover
- Blowgun
- Safety glasses
- Gloves
- Ear protection
- Solvent coatings
- Undercoating
- Paper wipes
- Tack cloths

Instructions

In the lab, vehicles have been prepared for use. Follow the work orders for each job and proceed as directed.

Vehicle Description

Year_____ Make _____ Model _____

VIN _____ Paint Code _____

Coating Used _____ Reduction Ratio _____

PROCEDURE

1. After reading the work order, gather the safety gear needed to complete the task. In the space provided below, list the personal and environmental safety equipment and precautions needed for this assignment. Have the instructor check and approve your plan before proceeding.

   _____

   _____

   _____

INSTRUCTOR'S APPROVAL _____

2. Calculate the amount of paint needed to blend the part provided. Mix and apply, using a standard blend technique.

   Notes:

   _____

   _____

   _____

3. Calculate the amount of paint needed to paint the part provided. Mix and apply, using reverse blending technique.

   Notes:

   _____

   _____

   _____

4. Calculate the amount of paint needed to paint the part provided. Mix and apply, using a wet-bed blending technique.

   Notes:

   _____

   _____

   _____

INSTRUCTOR COMMENTS:

_____

_____

_____

_____

_____

_____

WATERBORNE

Name _____   Date _____

Class _____   Instructor _____   Grade _____

OBJECTIVES

- Calculate the amount of coating needed for specific applications.
- Understand and be able to paint at high transfer efficiency level.
- Understand and perform techniques of application that will reduce time and material.
- Understand and be able to perform blending techniques such as:
  - Standard blending
  - Reverse blending
  - Wet-bedding blending
  - Zone blending
  - Single-stage blending
- Repair multistage paints.
- Understand waterborne application.

NATEF TASK CORRELATION

The written and hands-on activities in this chapter satisfy the NATEF High Priority-Individual and High Priority-Group requirements for Section IV: Paint and Refinishing, Subsections A. 1-6; C. 1-3; D. 1-10.

Tools and equipment needed (NATEF tool list)

- Pen and pencil
- Respirator
- Paint suit
- Coating technical data sheets (TDS)
- Waterborne coatings
- Wax and grease remover
- Blowgun
- Safety glasses
- Gloves
- Ear protection
- Solvent coatings
- Undercoating
- Paper wipes
- Tack cloths

Instructions

In the lab, vehicles have been prepared for use. Follow the work orders for each job and proceed as directed.

Vehicle Description

Year_____ Make _____ Model _____

VIN _____ Paint Code _____

Coating Used _____ Reduction Ratio _____

PROCEDURE

1. After reading the work order, gather the safety gear needed to complete the task. In the space provided below, list the personal and environmental safety equipment and precautions needed for this assignment. Have the instructor check and approve your plan before proceeding.

    _____

    _____

    _____

INSTRUCTOR'S APPROVAL _____

2. Calculate the amount of waterborne paint needed to blend the part provided. Mix and apply, using a standard blend technique.

    Notes:

    _____

    _____

    _____

3. Calculate the amount of waterborne paint needed to paint the part provided. Mix and apply, using the reverse blending technique.

    Notes:

    _____

    _____

    _____

4. Calculate the amount of waterborne paint needed to paint the part provided. Mix and apply, using a wet-bed blending technique.

    Notes:

    _____

    _____

    _____

INSTRUCTOR COMMENTS:

_____

_____

_____

_____

_____

_____

Name _____ Date _____

Class _____ Instructor _____ Grade _____

1. *Technician A* says that painters who have been painting for years can develop an eye for paint mixing and no longer need to measure when they reduce paint. *Technician B* says that paint reduction is critical, and paint must be measured each time it is mixed no matter what the painter's experience. Who is correct?
   A. Technician A only
   B. Technician B only
   C. Both Technicians A and B
   D. Neither Technician A nor B

2. *Technician A* says that mixing by scale is the most accurate method. *Technician B* says that a paint's pot life refers to how long it can stay in a can after it is shipped from the factory to the paint shop. Who is correct?
   A. Technician A only
   B. Technician B only
   C. Both Technicians A and B
   D. Neither Technician A nor B

3. *Technician A* says that a 2K product refers to epoxy paint products only. *Technician B* says that RTS stands for reactive trisulfide. Who is correct?
   A. Technician A only
   B. Technician B only
   C. Both Technicians A and B
   D. Neither Technician A nor B

4. *Technician A* says that in a paint ratio of 4-1-1, the 4 stands for the amount of reducer that is added to the mixture. *Technician B* says that a painter with a high transfer efficiency is a painter who will save the paint department money by using fewer materials. Who is correct?
   A. Technician A only
   B. Technician B only
   C. Both Technicians A and B
   D. Neither Technician A nor B

5. *Technician A* says that some undercoats can be efficiently brushed or rolled on to speed up production. *Technician B* says that engine bays are always the same color as the vehicle's outer color. Who is correct?
   A. Technician A only
   B. Technician B only
   C. Both Technicians A and B
   D. Neither Technician A nor B

6. *Technician A* says that edging-in of parts is often the paint apprentice's job. *Technician B* says that Wet-on-Wet sealer is a type of sealer that allows the color to be applied with only a short flash time, thus speeding up paintwork. Who is correct?
   A. Technician A only
   B. Technician B only
   C. Both Technicians A and B
   D. Neither Technician A nor B

7. *Technician A* says that color that was mixed for another car should never be used to edge in parts even when it is a close match. *Technician B* says that blending, though used occasionally in paint shops, is too costly to use on a regular basis. Who is correct?
   A. Technician A only
   B. Technician B only
   C. Both Technicians A and B
   D. Neither Technician A nor B

8. *Technician A* says that standard blending means a technician sprays the largest area coat first, with each coat thereafter being smaller. *Technician B* says that the wet-bedding blending spray technique is used for solid colors only. Who is correct?
   A. Technician A only
   B. Technician B only
   C. Both Technicians A and B
   D. Neither Technician A nor B

9. *Technician A* says that when preparing a panel for blending, one should sand the entire panel to P400/P500 or equivalent. *Technician B* says that when a panel is being prepared for blending, the area that will receive color is sanded to P400/P500 or equivalent. Who is correct?
   A. Technician A only
   B. Technician B only
   C. Both Technicians A and B
   D. Neither Technician A nor B

10. *Technician A* says that the when preparing a panel for blending, the area that will be cleared should be sanded with P1000 or equivalent. *Technician B* says that clear can be used over some single-stage blends. Who is correct?
    A. Technician A only
    B. Technician B only
    C. Both Technicians A and B
    D. Neither Technician A nor B

11. *Technician A* says that a multistage paint's midcoat could contain pearl, prismatic color, or even be tinted with color. *Technician B* says that a sprayout panel is made to find out how many midcoats are needed with a multistage paint. Who is correct?
    A. Technician A only
    B. Technician B only
    C. Both Technicians A and B
    D. Neither Technician A nor B

12. *Technician A* says that the ground coat of a multistage paint job should have a letdown panel made so the technician knows how many coats are needed. *Technician B* says that a multistage paint job will take up less room on a panel than a standard blend. Who is correct?
    A. Technician A only
    B. Technician B only
    C. Both Technicians A and B
    D. Neither Technician A nor B

13. *Technician A* says that because there are so many coats that must be applied to a multicoat paint job, the flash times recommended can be cut in half to speed up the work. *Technician B* says that both the ground coat and the midcoat need to be checked for color match before application, because either being off could cause the new finish to be an unacceptable match. Who is correct?
    A. Technician A only
    B. Technician B only
    C. Both Technicians A and B
    D. Neither Technician A nor B

14. *Technician A* says that the plan of attack for a complete paint job is not as critical as for a blend. *Technician B* says that the type of reducer and hardener used for a complete paint job may be one that causes the evaporation and curing to be slower than a blend. Who is correct?
    A. Technician A only
    B. Technician B only
    C. Both Technicians A and B
    D. Neither Technician A nor B

15. *Technician A* says that after blending the color into a panel, the clear should be applied to the entire panel. *Technician B* says that when blending clear, even if it looks good after it cures, it may fail later from exposure to the elements. Who is correct?
    A. Technician A only
    B. Technician B only
    C. Both Technicians A and B
    D. Neither Technician A nor B

Chapter 30

MUNSELL® COLOR TREE

Color Evaluation and Adjustment

■ WORK ASSIGNMENT 30-1

COLOR THEORY

Name _____ Date _____

Class _____ Instructor _____ Grade _____

NATEF TASKS A. 1-6; D. 1, 9-14

1. After reading the assignment, in the space provided below, list the personal and environmental safety equipment and precautions needed for this assignment.

   _____
   _____
   _____

2. List the primary colors. Record your findings:

   _____
   _____
   _____

3. Describe hue. Record your findings:

   _____
   _____
   _____

4. Describe value as it applies to color. Record your findings:

   _____
   _____
   _____

5. Describe chroma. Record your findings:

   _____
   _____

6. When should a technician tint? Record your findings:

   _____
   _____

7. How does light affect color? Record your findings:

   _____
   _____

8. Describe light refraction. Record your findings:

_____

_____

_____

9. Describe light reflection. Record your findings:

_____

_____

_____

10. What is metamerism? Record your findings:

_____

_____

_____

11. How can a technician avoid having his or her eyes become color fatigued? Record your findings:

_____

_____

_____

12. Why is it necessary to have color variance formulas? Record your findings:

_____

_____

_____

13. How can a toner become compromised, thus producing mismatches when used? Record your findings:

_____

_____

_____

14. Explain both metallic and micas. Record your findings:

_____

_____

_____

INSTRUCTOR COMMENTS:

_____

_____

_____

_____

_____

_____

■ WORK ASSIGNMENT 30-2

SPRAYOUT AND LETDOWN PANELS

Name _____ Date _____

Class _____ Instructor _____ Grade _____

NATEF TASKS A. 1-6; D. 1, 9-14

1. After reading the assignment, in the space provided below, list the personal and environmental safety equipment and precautions needed for this assignment.

   _____

   _____

   _____

2. Why should a technician make test panels? Record your findings:

   _____

   _____

3. What three major pieces of information will a sprayout panel tell a technician? Record your findings:

   _____

   _____

4. What is a color effect panel and how is it made? Record your findings:

   _____

   _____

5. If the TDS calls for two to three coats of color for coverage, how can a technician know for certain whether two or three coats are needed? Record your findings:

   _____

   _____

6. A metallic basecoat/clearcoat vehicle is to be painted, and the technician knows that the color is hard to match. Which test panel should the technician make, a letdown panel or a color effect panel? Why? Record your findings:

   _____

   _____

7. How is a letdown panel made? Record your findings:

   _____

   _____

8. Where would a technician find bare ground coat to which to compare the letdown panel? Record your findings:

   _____

INSTRUCTOR COMMENTS:

_____

_____

_____

_____

_____

_____

COLOR PLOTTING

Name _____   Date _____

Class _____   Instructor _____   Grade _____

NATEF TASKS A. 1-6; D. 1, 9-14

1. After reading the assignment, in the space provided below, list the personal and environmental safety equipment and precautions needed for this assignment.

_____

_____

_____

2. If as many as 50% of vehicles will need to have a color variant, what could cause this color variation? Record your findings:

_____

_____

_____

3. Suppose that the newly mixed color is sprayed out and the technician finds that the vehicle is bluer than the refinish color. Should the technician just get a blue shade off the tinting bank and add it to the refinish color? If not, why? Record your findings:

_____

_____

_____

4. Which of the three dimensions of color should a technician evaluate first? Record your findings:

_____

_____

_____

5. When should the technician evaluate chroma? Record your findings:

_____

_____

_____

6. How would a paint manufacturer's tinting chart be helpful in deciding on the correct color with which to tint? Record your findings:

_____

_____

_____

7. Why should the mixed paint be divided before tinting? Record your findings:

_____

_____

_____

INSTRUCTOR COMMENTS:

_____

_____

_____

_____

_____

_____

SPRAYOUT AND LETDOWN PANELS

Name _____   Date _____

Class _____   Instructor _____   Grade _____

OBJECTIVES

- Understand Munsell's color theory.
- Know the definitions and terms used to explain and understand color.
- Understand the effect of light on color.
- Understand how it reflects and how it refracts.
- Explain what metamerism is.
- Understand color deficiencies and how a technician can correct for them.
- Understand how color variances happen during manufacturing, how the paint manufacturers make adjustments for them, and what tools are at a painter's disposal to correct variances.
- Understand how pigments, metallic flakes, flop, mica, and prismatic colors affect a mixed and sprayed color.
- Understand why a color effect panel helps when tinting a mismatched color.
- Understand and perform the steps needed to color plot and tint a color, thus making it a blendable match.

NATEF TASK CORRELATION

The written and hands-on activities in this chapter satisfy the NATEF High Priority-Individual and High Priority-Group requirements for Section IV: Paint and Refinishing, A. 1-6; D. 1, 9-14.

Tools and equipment needed (NATEF tool list)

- Pen and pencil
- Respirator
- Paint suit
- Wax and grease remover
- Blowgun
- Mixing bank
- Color retrieval system
- Color-corrected light
- Safety glasses
- Gloves
- Ear protection
- Paper wipes
- Tack cloths
- Scale
- Variant system

Instructions

In the lab, vehicles have been prepared for use. Follow the work orders for each job and proceed as directed.

Vehicle Description

Year_____ Make _____ Model _____

VIN _____ Paint Code _____

Coating Used _____ Reduction Ratio _____

PROCEDURE

1. After reading the work order, gather the safety gear needed to complete the task. In the space provided below, list the personal and environmental safety equipment and precautions needed for this assignment. Have the instructor check and approve your plan before proceeding.

   _____

   _____

   _____

INSTRUCTOR'S APPROVAL _____

2. Using the color variant deck, evaluate the vehicle to determine if a nonprime color would be a better match. Record your findings:

   _____

   _____

   _____

3. Make a sprayout panel using the chosen color formula.

   Notes:

   _____

   _____

   _____

4. Is the color a blendable match?

   _____

   _____

   _____

5. How many coats were needed to reach hiding?

   _____

   _____

   _____

6. Should the formula be tinted?

   _____

   _____

   _____

7. Compare the sprayout panel to the vehicle. Will you need to alter the spray technique to obtain a blendable match?

   Notes:

   _____

   _____

   _____

8. Make a letdown panel for a tricoat.

   Notes:

   _____

   _____

   _____

9. Compare the letdown panel to the standard to determine if the ground coat matches and how many intermediate coats will need to be applied for a match. Record your findings:

   _____

   _____

INSTRUCTOR COMMENTS:

_____

_____

_____

_____

_____

_____

■ WORK ORDER 30-2

COLOR PLOTTING

Name _____ Date _____

Class _____ Instructor _____ Grade _____

OBJECTIVES

- Understand Munsell's color theory.
- Know the definitions and terms used to explain and understand color.
- Understand the effect of light on color.
- Understand how it reflects and how it refracts.
- Explain what metamerism is.
- Understand color deficiencies and how a technician can correct for them.
- Understand how color variances happen during manufacturing, how the paint manufacturers make adjustments for them, and what tools are at a painter's disposal to correct variances.
- Understand how pigments, metallic flakes, flop, mica, and prismatic colors affect a mixed and sprayed color.
- Understand why a color effect panel helps when tinting a mismatched color.
- Understand and perform the steps needed to color plot and tint a color, thus making it a blendable match.

NATEF TASK CORRELATION

The written and hands-on activities in this chapter satisfy the NATEF High Priority-Individual and High Priority-Group requirements for Section IV: Paint and Refinishing, A. 1-6; D. 1, 9-14.

Tools and equipment needed (NATEF tool list)

- Pen and pencil
- Respirator
- Paint suit
- Wax and grease remover
- Blowgun
- Mixing bank
- Color retrieval system
- Color-corrected light

- Safety glasses
- Gloves
- Ear protection
- Paper wipes
- Tack cloths
- Scale
- Variant system

Instructions

In the lab, vehicles have been prepared for use. Follow the work orders for each job and proceed as directed.

Vehicle Description

Year_____ Make _____ Model _____

VIN _____ Paint Code _____

Coating Used _____ Reduction Ratio _____

PROCEDURE

1. After reading the work order, gather the safety gear needed to complete the task. In the space provided below, list the personal and environmental safety equipment and precautions needed for this assignment. Have the instructor check and approve your plan before proceeding.

_____

_____

_____

INSTRUCTOR'S APPROVAL _____

2. On the prepared vehicle, plot the nonmetallic color for tinting.

   Notes:

   _____

   _____

   _____

3. Is the vehicle lighter or darker than the mixed paint?

   Notes:

   _____

   _____

   _____

4. Is the vehicle redder, bluer, or yellower than the mixed paint?

   Notes:

   _____

   _____

   _____

5. Using the tinting chart, determine which color will move the mixed formula in the correct direction.

   Notes:

   _____

   _____

   _____

6. Make a sprayout panel of the tinted formula and compare it to the vehicle.

   Notes:

   _____

   _____

   _____

7. Apply the tinted color to the vehicle and evaluate.

   Notes:

   _____

   _____

   _____

INSTRUCTOR COMMENTS:

_____

_____

_____

_____

_____

1. *Technician A* says that ROY G BIV is a mnemonic (memory device) for red, orange, yellow, green, blue, indigo, and violet. *Technician B* says that primary colors are yellow, blue, red, and green. Who is correct?
   A. Technician A only
   B. Technician B only
   C. Both Technicians A and B
   D. Neither Technician A nor B

2. *Technician A* says that Albert H. Munsell developed a test for color deficiency in 1917. *Technician B* says that Albert H. Munsell developed a three-dimensional color tree used to identify and describe color. Who is correct?
   A. Technician A only
   B. Technician B only
   C. Both Technicians A and B
   D. Neither Technician A nor B

3. *Technician A* says that hue is the dimension of color that describes its color. *Technician B* says that chroma is the dimension of color that describes its lightness or darkness. Who is correct?
   A. Technician A only
   B. Technician B only
   C. Both Technicians A and B
   D. Neither Technician A nor B

4. *Technician A* says that value describes how intense a color is. *Technician B* says that chroma is sometimes called intensity. Who is correct?
   A. Technician A only
   B. Technician B only
   C. Both Technicians A and B
   D. Neither Technician A nor B

5. *Technician A* says that if a color is found to be a nonblendable color, the first consideration should be tinting. *Technician B* says that there are few things that can cause a mismatch other than the tints added to a formula. Who is correct?
   A. Technician A only
   B. Technician B only
   C. Both Technicians A and B
   D. Neither Technician A nor B

6. *Technician A* says that factory color codes are found on the driver's door of vehicles. *Technician B* says that to locate where on a vehicle the manufacturer's paint code is found, one should check the vehicle's collision repair manual. Who is correct?
   A. Technician A only
   B. Technician B only
   C. Both Technicians A and B
   D. Neither Technician A nor B

7. *Technician A* says that color can only move either right or left of itself on a color wheel when describing a color. *Technician B* says that a vehicle viewed under fluorescent light will look the same when it is compared to sunlight. Who is correct?
   A. Technician A only
   B. Technician B only
   C. Both Technicians A and B
   D. Neither Technician A nor B

8. *Technician A* says that CRI is a measurement of color-corrected light. *Technician B* says that CRI stands for color rendering index. Who is correct?
   A. Technician A only
   B. Technician B only
   C. Both Technicians A and B
   D. Neither Technician A nor B

9. *Technician A* says that reflection means a light passes through something and a different color comes out the other side. *Technician B* says that reflection means light bounces off a surface and a color is viewed. Who is correct?
    A. Technician A only
    B. Technician B only
    C. Both Technicians A and B
    D. Neither Technician A nor B

10. *Technician A* says that metamerism is a condition in which a color appears different under different types of light. *Technician B* says that flop is a condition in which a color containing metallic and/or mica looks different at different angles. Who is correct?
    A. Technician A only
    B. Technician B only
    C. Both Technicians A and B
    D. Neither Technician A nor B

11. *Technician A* says that a side tone angle means a vehicle's color is evaluated at a 45- to 60-degree angle. *Technician B* says that metamerism should be judged by viewing it at a side tone angle. Who is correct?
    A. Technician A only
    B. Technician B only
    C. Both Technicians A and B
    D. Neither Technician A nor B

12. *Technician A* says that color deficiency means a person sees no color at all. *Technician B* says that color deficiency affects women more than men. Who is correct?
    A. Technician A only
    B. Technician B only
    C. Both Technicians A and B
    D. Neither Technician A nor B

13. *Technician A* says that color fatigue can occur after staring at a color for only 5 seconds. *Technician B* says that a color variance on a vehicle can occur because a model of a car could be painted at different factories. Who is correct?
    A. Technician A only
    B. Technician B only
    C. Both Technicians A and B
    D. Neither Technician A nor B

14. *Technician A* says that "Chromalusion" is a brand of prismatic color. *Technician B* says that a sprayout or color effect panel is a test panel used to check a multistage color. Who is correct?
    A. Technician A only
    B. Technician B only
    C. Both Technicians A and B
    D. Neither Technician A nor B

15. *Technician A* says that to know the number of coats needed for a multistage color, a sprayout panel should be made. *Technician B* says that when plotting a color, it should be evaluated in the order of value, hue, and chroma. Who is correct?
    A. Technician A only
    B. Technician B only
    C. Both Technicians A and B
    D. Neither Technician A nor B

Chapter 31

Paint Problems and Prevention

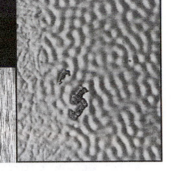

DEFECTS CAUSED BY POOR PREPARATION

Name _____   Date _____

Class _____   Instructor _____   Grade _____

NATEF TASKS A. 1-6; E. 1-28

1. After reading the assignment, in the space provided below, list the personal and environmental safety equipment and precautions needed for this assignment.

_____

_____

_____

2. Describe how to identify dirt and explain its cause.

_____

_____

_____

3. Explain how to prevent it:

_____

_____

_____

4. Explain how to repair it:

_____

_____

5. Describe how to identify sand scratch swelling and explain its cause.

_____

_____

_____

6. Explain how to prevent it:

_____

_____

_____

7. Explain how to repair it:

_____

_____

_____

8. Describe how to identify overspray and explain its cause.

_____

_____

_____

9. Explain how to prevent it:

_____

_____

_____

10. Explain how to repair it:

_____

_____

_____

11. Describe how to identify bleeding and explain its cause.

_____

_____

_____

12. Explain how to prevent it:

_____

_____

_____

13. Explain how to repair it:

_____

_____

_____

14. Describe how to identify contour mapping and explain its cause.

_____

_____

_____

15. Explain how to prevent it:

_____

_____

_____

16. Explain how to repair it:

_____

_____

_____

17. Describe how to identify lifting and explain its cause.

_____

_____

_____

18. Explain how to prevent it:

_____

_____

_____

19. Explain how to repair it:

_____

_____

_____

20. Describe how to identify peeling and explain its cause.

_____

_____

_____

21. Explain how to prevent it:

_____

_____

_____

22. Explain how to repair it:

_____

_____

_____

23. Describe how to identify chips and explain their cause.

_____

_____

_____

24. Explain how to prevent them:

_____

_____

_____

25. Explain how to repair them:

_____

_____

_____

INSTRUCTOR COMMENTS:

_____

_____

_____

_____

_____

DEFECTS CAUSED BY POOR SPRAY TECHNIQUES

Name _____ Date _____

Class _____ Instructor _____ Grade _____

1. After reading the assignment, in the space provided below, list the personal and environmental safety equipment and precautions needed for this assignment.

   _____

   _____

   _____

2. Describe how to identify runs or sags and explain the cause.

   _____

   _____

   _____

3. Explain how to prevent them:

   _____

   _____

   _____

4. Explain how to repair them:

   _____

   _____

   _____

5. Describe how to identify orange peel and explain its cause.

   _____

   _____

   _____

6. Explain how to prevent it:

   _____

   _____

   _____

7. Explain how to repair it:

   _____

   _____

   _____

8. Describe how to identify striping and explain its cause.

   _____

   _____

   _____

9. Explain how to prevent it:

   _____

   _____

   _____

10. Explain how to repair it:

_____

_____

_____

11. Describe how to identify poor hiding and explain its cause.

_____

_____

_____

12. Explain how to prevent it:

_____

_____

_____

13. Explain how to repair it:

_____

_____

_____

14. Describe how to identify solvent popping and explain its cause.

_____

_____

_____

15. Explain how to prevent it:

_____

_____

_____

16. Explain how to repair it:

_____

_____

_____

17. Describe how to identify mottling and explain its cause.

_____

_____

_____

18. Explain how to prevent it:

_____

_____

_____

19. Explain how to repair it:

_____

_____

_____

20. Describe how to identify dry spray and explain its cause.

_____

_____

_____

21. Explain how to prevent it:

_____

_____

_____

22. Explain how to repair it:

_____

_____

_____

INSTRUCTOR COMMENTS:

_____

_____

_____

_____

_____

_____

DEFECTS CAUSED BY CONTAMINATION

Name _____ Date _____

Class _____ Instructor _____ Grade _____

1. After reading the assignment, in the space provided below, list the personal and environmental safety equipment and precautions needed for this assignment.

   _____

   _____

   _____

2. Describe how to identify fish eye and explain its cause.

   _____

   _____

   _____

3. Explain how to prevent it:

   _____

   _____

   _____

4. Explain how to repair it:

   _____

   _____

   _____

5. Describe how to identify blistering and explain its cause.

   _____

   _____

   _____

6. Explain how to prevent it:

   _____

   _____

   _____

7. Explain how to repair it:

   _____

   _____

   _____

8. Describe how to identify acid rain and explain its cause.

   _____

   _____

   _____

9. Explain how to prevent it:

   _____

   _____

   _____

10. Explain how to repair it:

_____

_____

_____

11. Describe how to identify bird droppings and explain the cause.

_____

_____

_____

12. Explain how to prevent them:

_____

_____

_____

13. Explain how to repair them:

_____

_____

_____

INSTRUCTOR COMMENTS:

_____

_____

_____

_____

_____

_____

DEFECTS CAUSED BY POOR DRYING OR CURING

Name _____ Date _____

Class _____ Instructor _____ Grade _____

1. After reading the assignment, in the space provided below, list the personal and environmental safety equipment and precautions needed for this assignment.

   _____

   _____

   _____

2. Describe how to identify dieback and explain its cause.

   _____

   _____

   _____

3. Explain how to prevent it:

   _____

   _____

   _____

4. Explain how to repair it:

   _____

   _____

   _____

5. Describe how to identify cracking and explain its cause.

   _____

   _____

   _____

6. Explain how to prevent it:

   _____

   _____

   _____

7. Explain how to repair it:

   _____

   _____

   _____

8. Describe how to identify wrinkling and explain its cause.

   _____

   _____

   _____

9. Explain how to prevent it:

   _____

   _____

   _____

10. Explain how to repair it:

_____
_____
_____

INSTRUCTOR COMMENTS:

_____
_____
_____
_____
_____
_____

DEFECTS CAUSED BY POOR PREPARATION

Name _____   Date _____

Class _____   Instructor _____   Grade _____

OBJECTIVES

- Be able to identify, understand the cause, prevent the occurrence, and repair:
  - Defects caused by poor preparation
  - Defects caused by poor spray techniques
  - Defects caused by contamination
  - Defects caused by poor drying or curing

NATEF TASK CORRELATION

The written and hands-on activities in this chapter satisfy the NATEF High Priority-Individual and High Priority-Group requirements for Section IV: Paint and Refinishing, Subsections A. 1-6; E. 1-28.

Tools and equipment needed (NATEF tool list)

- Pen and pencil
- Respirator
- Lighted magnifying glass
- Paper wipes
- Polishing compounds
- Nibbing tools

- Safety glasses
- Gloves
- Wax and grease remover
- Buffers
- Assorted sandpaper
- Color corrected light

Instructions

In the lab, vehicles have been prepared for use. Follow the work orders for each job and proceed as directed.

Vehicle Description

Year_____ Make _____ Model _____

VIN _____ Paint Code _____

Coating Used _____ Reduction Ratio _____

PROCEDURE

1. After reading the work order, gather the safety gear needed to complete the task. In the space provided below, list the personal and environmental safety equipment and precautions needed for this assignment. Have the instructor check and approve your plan before proceeding.

_____

_____

_____

INSTRUCTOR'S APPROVAL _____

On the vehicle or part provided:

2. Identify dirt and list its possible causes and what could have been done to prevent it.

_____

_____

_____

Remove the defect.

Notes:

_____

_____

_____

If the defect cannot be removed by detailing, list the repair procedure for its correction.

_____

_____

_____

3. Identify sand scratch swelling and list its possible causes and what could have been done to prevent it.

_____

_____

_____

Remove the defect.

Notes:

_____

_____

_____

If the defect cannot be removed by detailing, list the repair procedure for its correction.

_____

_____

_____

4. Identify overspray and list its possible causes and what could have been done to prevent it.

_____

_____

_____

Remove the defect.

Notes:

_____

_____

_____

If the defect cannot be removed by detailing, list the repair procedure for its correction.

_____

_____

_____

5. Identify bleeding and list its possible causes and what could have been done to prevent it.

_____

_____

_____

Remove the defect.

Notes:

_____

_____

_____

If the defect cannot be removed by detailing, list the repair procedure for its correction.

_____

_____

_____

6. Identify contour mapping and list its possible causes and what could have been done to prevent it.

_____

_____

_____

Remove the defect.

Notes:

_____

_____

_____

If the defect cannot be removed by detailing, list the repair procedure for its correction.

_____

_____

_____

7. Identify lifting and list its possible causes and what could have been done to prevent it.

_____

_____

_____

Remove the defect.

Notes:

_____

_____

_____

If the defect cannot be removed by detailing, list the repair procedure for its correction.

_____

_____

_____

8. Identify peeling and list its possible causes and what could have been done to prevent it.

_____

_____

_____

Remove the defect.

Notes:

_____

_____

_____

If the defect cannot be removed by detailing, list the repair procedure for its correction.

_____

_____

_____

9. Identify chips and list the possible causes and what could have been done to prevent them.

_____

_____

_____

Remove the defect.

Notes:

_____

_____

_____

If the defect cannot be removed by detailing, list the repair procedure for its correction.

_____

_____

_____

INSTRUCTOR COMMENTS:

_____

_____

_____

_____

_____

_____

DEFECTS CAUSED BY POOR SPRAY TECHNIQUES

Name _____ Date _____

Class _____ Instructor _____ Grade _____

OBJECTIVES

- Be able to identify, understand the cause, prevent the occurrence, and repair:
  - Defects caused by poor preparation
  - Defects caused by poor spray techniques
  - Defects caused by contamination
  - Defects caused by poor drying or curing

NATEF TASK CORRELATION

The written and hands-on activities in this chapter satisfy the NATEF High Priority-Individual and High Priority-Group requirements for Section IV: Paint and Refinishing, Subsections A. 1-6; E. 1-28.

Tools and equipment needed (NATEF tool list)

- Pen and pencil
- Respirator
- Lighted magnifying glass
- Paper wipes
- Polishing compounds
- Nibbing tools
- Safety glasses
- Gloves
- Wax and grease remover
- Buffers
- Assorted sandpaper
- Color corrected light

Instructions

In the lab, vehicles have been prepared for use. Follow the work orders for each job and proceed as directed.

Vehicle Description

Year_____ Make _____ Model _____

VIN _____ Paint Code _____

Coating Used _____ Reduction Ratio _____

PROCEDURE

1. After reading the work order, gather the safety gear needed to complete the task. In the space provided below, list the personal and environmental safety equipment and precautions needed for this assignment. Have the instructor check and approve your plan before proceeding.

_____

_____

_____

INSTRUCTOR'S APPROVAL _____

On the vehicle or part provided:

2. Identify runs or sags and list the possible causes and what could have been done to prevent them.

_____

_____

_____

Remove the defect.

Notes:

_____

_____

_____

If the defect cannot be removed by detailing, list the repair procedure for its correction.

_____

_____

_____

3. Identify orange peel and list its possible causes and what could have been done to prevent it.

_____

_____

_____

Remove the defect.

Notes:

_____

_____

_____

If the defect cannot be removed by detailing, list the repair procedure for its correction.

_____

_____

_____

4. Identify striping and list its possible causes and what could have been done to prevent it.

_____

_____

_____

Remove the defect.

Notes:

_____

_____

_____

If the defect cannot be removed by detailing, list the repair procedure for its correction.

_____

_____

_____

5. Identify poor hiding and list its possible causes and what could have been done to prevent it.

_____

_____

_____

Remove the defect.
Notes:

_____

_____

_____

If the defect cannot be removed by detailing, list the repair procedure for its correction.

_____

_____

_____

6. Identify solvent popping and list its possible causes and what could have been done to prevent it.

_____

_____

_____

Remove the defect.
Notes:

_____

_____

_____

If the defect cannot be removed by detailing, list the repair procedure for its correction.

_____

_____

_____

7. Identify mottling and list its possible causes and what could have been done to prevent it.

_____

_____

_____

Remove the defect.
Notes:

_____

_____

_____

If the defect cannot be removed by detailing, list the repair procedure for its correction.

_____

_____

_____

8. Identify dry spray and list its possible causes and what could have been done to prevent it.

_____

_____

_____

Remove the defect.
Notes:

_____

_____

_____

If the defect cannot be removed by detailing, list the repair procedure for its correction.

_____

_____

_____

INSTRUCTOR COMMENTS:

_____

_____

_____

_____

_____

_____

DEFECTS CAUSED BY CONTAMINATION

Name _____   Date _____

Class _____   Instructor _____   Grade _____

OBJECTIVES

- Be able to identify, understand the cause, prevent the occurrence, and repair:
  - Defects caused by poor preparation
  - Defects caused by poor spray techniques
  - Defects caused by contamination
  - Defects caused by poor drying or curing

NATEF TASK CORRELATION

The written and hands-on activities in this chapter satisfy the NATEF High Priority-Individual and High Priority-Group requirements for Section IV: Paint and Refinishing, Subsections A. 1-6; E. 1-28.

Tools and equipment needed (NATEF tool list)

- Pen and pencil
- Respirator
- Lighted magnifying glass
- Paper wipes
- Polishing compounds
- Nibbing tools
- Safety glasses
- Gloves
- Wax and grease remover
- Buffers
- Assorted sandpaper
- Color corrected light

Instructions

In the lab, vehicles have been prepared for use. Follow the work orders for each job and proceed as directed.

Vehicle Description

Year_____ Make _____ Model _____

VIN _____ Paint Code _____

Coating Used _____ Reduction Ratio _____

PROCEDURE

1. After reading the work order, gather the safety gear needed to complete the task. In the space provided below, list the personal and environmental safety equipment and precautions needed for this assignment. Have the instructor check and approve your plan before proceeding.

_____

_____

_____

INSTRUCTOR'S APPROVAL _____

On the vehicle or part provided:

2. Identify fish eye and list its possible causes and what could have been done to prevent it.

_____

_____

_____

Remove the defect.
Notes:

_____

_____

_____

If the defect cannot be removed by detailing, list the repair procedure for its correction.

_____

_____

_____

3. Identify blistering and list its possible causes and what could have been done to prevent it.

_____

_____

_____

Remove the defect.
Notes:

_____

_____

_____

If the defect cannot be removed by detailing, list the repair procedure for its correction.

_____

_____

_____

4. Identify acid rain and list its possible causes and what could have been done to prevent it.

_____

_____

_____

Remove the defect.
Notes:

_____

_____

_____

If the defect cannot be removed by detailing, list the repair procedure for its correction.

_____

_____

_____

5. Identify bird droppings and list the possible causes and what could have been done to prevent them.

_____

_____

_____

Remove the defect.

Notes:

_____

_____

_____

If the defect cannot be removed by detailing, list the repair procedure for its correction.

_____

_____

6. Identify solvent popping and list its possible causes and what could have been done to prevent it.

_____

_____

Remove the defect.

Notes:

_____

_____

_____

If the defect cannot be removed by detailing, list the repair procedure for its correction.

_____

_____

7. Identify mottling and list its possible causes and what could have been done to prevent it.

_____

_____

_____

Remove the defect.

Notes:

_____

_____

_____

If the defect cannot be removed by detailing, list the repair procedure for its correction.

_____

_____

8. Identify dry spray and list its possible causes and what could have been done to prevent it.

_____

_____

_____

Remove the defect.

Notes:

_____

_____

_____

If the defect cannot be removed by detailing, list the repair procedure for its correction.

_____

_____

_____

INSTRUCTOR COMMENTS:

_____

_____

_____

_____

_____

_____

■ WORK ORDER 31-4

DEFECTS CAUSED BY POOR DRYING OR CURING

Name _____   Date _____

Class _____   Instructor _____   Grade _____

OBJECTIVES

- Be able to identify, understand the cause, prevent the occurrence, and repair:
  - Defects caused by poor preparation
  - Defects caused by poor spray techniques
  - Defects caused by contamination
  - Defects caused by poor drying or curing

NATEF TASK CORRELATION

The written and hands-on activities in this chapter satisfy the NATEF High Priority-Individual and High Priority-Group requirements for Section IV: Paint and Refinishing, Subsections A. 1-6; E. 1-28.

Tools and equipment needed (NATEF tool list)

- Pen and pencil
- Respirator
- Lighted magnifying glass
- Paper wipes
- Polishing compounds
- Nibbing tools
- Safety glasses
- Gloves
- Wax and grease remover
- Buffers
- Assorted sandpaper
- Color corrected light

Instructions

In the lab, vehicles have been prepared for use. Follow the work orders for each job and proceed as directed.

Vehicle Description

Year_____ Make _____ Model _____

VIN _____ Paint Code _____

Coating Used _____ Reduction Ratio _____

PROCEDURE

1. After reading the work order, gather the safety gear needed to complete the task. In the space provided below, list the personal and environmental safety equipment and precautions needed for this assignment. Have the instructor check and approve your plan before proceeding.

_____

_____

_____

INSTRUCTOR'S APPROVAL _____

On the vehicle or part provided:

2. Identify dieback and list its possible causes and what could have been done to prevent it.

_____

_____

_____

Remove the defect.

Notes:

_____

_____

If the defect cannot be removed by detailing, list the repair procedure for its correction.

_____

_____

_____

3. Identify cracking and list its possible causes and what could have been done to prevent it.

_____

_____

_____

Remove the defect.

Notes:

_____

_____

_____

If the defect cannot be removed by detailing, list the repair procedure for its correction.

_____

_____

_____

4. Identify wrinkling and list its possible causes and what could have been done to prevent it.

_____

_____

_____

Remove the defect.

Notes:

_____

_____

_____

If the defect cannot be removed by detailing, list the repair procedure for its correction.

_____

_____

_____

INSTRUCTOR COMMENTS:

_____

_____

_____

_____

_____

Name _____   Date _____

Class _____   Instructor _____   Grade _____

1. *Technician A* says that dirt contamination occurs from not properly cleaning the vehicle before it is painted. *Technician B* says that dirt contamination can sometimes be mistaken for solvent popping, but on close examination dirt is found to be more random in its distribution than solvent popping. Who is correct?
   A. Technician A only
   B. Technician B only
   C. Both Technicians A and B
   D. Neither Technician A nor B

2. *Technician A* says that sand scratch swelling is a defect that occurs due to airborne contamination. *Technician B* says that sand scratch swelling occurs due to poor surface preparation. Who is correct?
   A. Technician A only
   B. Technician B only
   C. Both Technicians A and B
   D. Neither Technician A nor B

3. *Technician A* says that contour mapping appears as sand scratches or splitting around the repair area. *Technician B* says that contour mapping is a condition in which the vehicle finish flakes off around areas that were not sanded properly. Who is correct?
   A. Technician A only
   B. Technician B only
   C. Both Technicians A and B
   D. Neither Technician A nor B

4. *Technician A* says that lifting occurs due to poor spray technique. *Technician B* says that poor curing and drying cause lifting. Who is correct?
   A. Technician A only
   B. Technician B only
   C. Both Technicians A and B
   D. Neither Technician A nor B

5. *Technician A* says that runs or sags are caused by poor surface preparation. *Technician B* says that poor spraying with too slow a travel speed causes runs and sags. Who is correct?
   A. Technician A only
   B. Technician B only
   C. Both Technicians A and B
   D. Neither Technician A nor B

6. *Technician A* says that if a gun is not adjusted properly, orange peel may occur. *Technician B* says that if too fast a gun travel speed is used, orange peel will occur. Who is correct?
   A. Technician A only
   B. Technician B only
   C. Both Technicians A and B
   D. Neither Technician A nor B

7. *Technician A* says that striping can be caused by poor overlap technique. *Technician B* says that striping can be caused by a dirty spray gun. Who is correct?
   A. Technician A only
   B. Technician B only
   C. Both Technicians A and B
   D. Neither Technician A nor B

8. *Technician A* says that solvent popping is caused by poor finish mixing. *Technician B* says that solvent popping is caused by improper flash time between coats. Who is correct?
   A. Technician A only
   B. Technician B only
   C. Both Technicians A and B
   D. Neither Technician A nor B

9. *Technician A* says that fish eye is caused by surface contamination before painting. *Technician B* says that dieback is caused by overreduction of the finish. Who is correct?
   A. Technician A only
   B. Technician B only
   C. Both Technicians A and B
   D. Neither Technician A nor B

10. *Technician A* says that cracking can be prevented by applying the recommended film thickness. *Technician B* says that wrinkling is cured by removing all cracks in the affected area, preparing the surface with undercoats and sealers according to the manufacturer's recommendations, and reapplying topcoat. Who is correct?
    A. Technician A only
    B. Technician B only
    C. Both Technicians A and B
    D. Neither Technician A nor B